AF559963

ANIMAL HUSBANDRY

ANIMAL HUSBANDRY

By

Ashok Kumar

Lecturer

Department of Zoology

University Campus Department

Jhansi

DPH

DISCOVERY PUBLISHING HOUSE PVT. LTD.

NEW DELHI-110 002

Reprinted - 2019

First Published - 2006

ISBN: 978-81-8356-072-6

Animal Husbandry

Published by:

DISCOVERY PUBLISHING HOUSE PVT. LTD.
4383/4B, Ansari Road, Darya Ganj
New Delhi-110 002 (India)
Phone: +91-11-23279245, 23253475; 43596065
E-mail: discoverybooksindia@gmail.com
discoverypublishinghouse@gmail.com
web: www.discoverypublishinggroup.com

Printed at:
Infinity Imaging Systems
Delhi

Preface

Animal Husbandry is intended primarily as a text for the basic course in Animal Industries and Farm Animals, as the case may be. It should prove useful as a reference book for those involved in livestock production. This book is not an encyclopaedia. Rather, it is a concise and up-to-date collection of those facts and judgements that can give a beginning student a well-rounded and clear picture of the scope, importance, and operation of our great livestock, packing, and diary industries. Since many students major in other fields, and make this their principal contact with a very important field of knowledge, an effort has been made to confine the material to the essential facts and to maintain student interest by the liberal use of pictures, tables and charts.

Purpose of the author has been to present a contemporary outline of the main subject divisions of animal husbandry with emphasis being placed upon the practical implications of each.

To make the work more comprehensive and informative, the author has consulted many authoritative books, research journals, abstracts, monographs etc. He is grateful to all those great scholars whose work are cited or substantially reproduced.

There can be no claim to originality except in the manner of treatment and much of the information has been obtained from the books and scientific journals available in the different libraries.

The author expresses his thanks to his friends and colleagues whose continues inspirations have initiated him to bring out this book.

The author is painfully aware of the shortcomings, errors and misprints that have crept in, and shall be grateful to receive suggestions for improvement of the next edition from all the readers.

The author expresses his gratitude to Mr. Wasan and staff of M/s Discovery Publishing House for their whole hearted co-operation in the publication of this book.

Author

CONTENTS

1. **Introduction** 1—27

Signs of Health, Horses, Cattle, Sheep, Pigs, Dogs, Cats, Poultry, Handling, Restraining Animals in Buildings, Headstall or Head Collars, Twitches, Gags, Cattle, Dogs, Casting Animals, Precautions before Casting, Administration of Medicine or Drenching, Horses and Cattle, Pigs, Sheep, Dogs, Cats, Birds, Injections, Subcutaneous Injections, Intramuscular Injections, Intraperitoneal Injections, Intravenous Injections, Intramammary Injection, Intra-Uterine Injections, Intra-Vaginal Tampons or Pessaries.

2. **Housing and Yarding** 28—68

Size, Requirements for Housing, Roof, Walls, Floors, Horses: in Temperate Climates, Loose Box, Stalls, Sick Horse Box, Feed Room and Barn, Saddle or Tack Room, Hot Climate Housing, Control of Flies, Fire Precautions, Cattle, Dairy Cows in Temperate Climates, Cowhouse, Milking Parlour, Housing and Yards, Feeding, Movement of Cows, Collecting Yards, Race and Crush, Dairy Cows in a Tropical Climate, Bull Pen, Calves, Beef Cattle, Beef Cattle Sheds, Slats, Bull Beef, Sheep, Sheep Handling and Dipping Yards, Pigs, Sow, Dry Sow Housing, Farrowing Accommodation, Weaning and Rearing, Fattening House, Boar, Poultry, Rearing Layers and Breeders, Broiler House, Tropical Poultry Houses, Turkeys, Breeding and Rearing Houses, Breeding, Rearing, Fattening, Ducks, Rabbits.

3. **Feeding** 69—135

Energy, Proteins, Minerals, Vitamins, Digestion, Monogastric Animals: Pig, Dog, Cat, Horse, Ruminants, Feeding Horses, Dairy Cows, Dry Matter Intake, Proteins, Developments in Practical Feeding of Dairy Cows, Minerals, Feeding Beef Cattle, Feeding Calves, Systems for Calf Feeding, Feeding after Weaning, Feeding Sheep, Mating, Pregnancy, Lactation, Importance of Protein Quality in Pregnancy, Feeding Lambs and Hoggs, Use of Grassland, Feeding Pigs, Practical Feeding, Creep Feeds, Grower and Finishing Rations, Pregnant Sow Ration, Lactating Sows, Feeding Boars, Feeding Poultry, Fowl, Broilers, Pullets and Broiler Breeders, Layers, Turkeys, Ducks, Rabbits, Dogs, Maintenance Requirements, Pregnant Bitch, Feeding during Lactation, Feeding Puppies, Cats, Adult Cats, Young Cats, Water, Water Requirements.

4. **Breeding** 136—157

Historical Outline: The Animal Improvers, Emergence of Modern Breeds, Scientific Basis of Stock Breeding, Cell Genetics, Mendelism, Effect of the Environment, Dairy Cattle in the Tropics, Breeding Programmes, Methods used to Produce Improved Strains of Animals, Inbreeding, Mass Selection, Recessive Characters, Cross-Breeding in Practice, Fat-Lamb Production, Cross-Breeding in Dairy Cattle, Beef Production, Pig Production, Poultry Production, General.

5. **Development** 158—182

Moving Towards Intensivism, Dressing Percentages, Sexual Development, Cattle, Birth of the Calf, Sheep, Birth of the Lamb, Pigs, Rabbits, General Growth Pattern, Rate of Growth and Carcass Composition, Growth in Pigs, Growth in Cattle, Factors Controlling Carcass Composition, Muscle Development, Influence of Breed on Muscle Growth, The Butcher and the Carcass, Comparative Studies of Muscle Growth, Carcass Assessment.

6. **Reproduction** 183—221

Comparative Physiology of Reproductive Processes, Male, Testes, Penis, Disorders of Reproduction in the Male, Method of Examination, Artificial Insemination, Collection of Semen, Evaluation of Semen, Dilution of Semen, Deep-frozen Bull Semen, Insemination, Mares, Cows, Sheep, Sows, Female, Ovary, Oestrous

Cycle, Pregnancy, Factors Controlling Fertility, Control of Oestrus by Hormones, Detection of Pregnancy, Parturition, Normal Parturition, Care of the Newborn, Disorders of Reproduction in the Female, Infection Following Parturition, Infections During Pregnancy, Other Causes of Infertility, Functional Infertility, Infertility and Nutrition, Production Disease.

7. **Piggery** 222—230

The Country Pig, Advantages of pig Production, Efficiency in Park Production, Selection of Breeds, Berkshire, American Breed, Chester White, Duroc, Hampshire, Feeding and Management of the Herd, Gestation Period, Artificial Insemination, Growth of Young, Proportional Growth of the Body, Slaughter of Pigs, Products of Piggery, Pork, Bristles, Sausages, Lard, Other Uses, Diseases and Control.

8. **Buffalo** 231—245

Breeds of Indian Buffaloes, Murrah, Nili/Ravi, Surti, Jaffarabadi, Mehsana, Uttar Pradesh Group, Bhadawari, Tarai, Central India Group, Nagpuri, Pandhepuri, Manda, Sambalpur, Kalhandi, Jerangi, South India Group, Toda, Breeding and Management Problems, Buffaloes are Seasonal Breeders, Silent Heat, Corrective Measures, Repeat Breeding, Weaning, Calf Mortality, Housing, Feeding Colostrum, Feeding Antibiotics, Feeding Milk Replacer, Disease.

9. **Horses and Mules** 246—255

Ideal Draft Animal, Size, Head and Neck, Shoulders, Body, Rear Quarters, Feet, Legs, and Action, Common Blemishes and Unsoundnesses, Market Classes and Grades of Horses and Mules, Draft Horses, Chunks, Wagon Horses, Driving Horses, Saddle Horses, Ponies, Classes and Grades of Mules, Factors Affecting Marketing Value of Horses and Mules.

10. **Preventive Medicine** 256—281

Dairy Cattle, Call-Rearing, Rearing of Yearling Heifers, Mating of Heifers, Cows, Diseases from Infections, Production Diseases, Biochemical Aids in Preventive Medicine, Reproduction Disorders, Importance of Regular Visits by Veterinarian, Improving Conception Rates, Pregnancy Diagnosis, Records, Beef Cattle, Controlled Mating, Sheep, Infections, Worms, Foot-rot, Pigs, Starting a Herd, Keeping Free of Infection, Preventing Perinatal Loss: Night

Farrowing, Worms in Pigs, Organization of Preventive Medicine for Pig Units, Vices, Poultry, Environment and Nutrition, Virus Diseases, Bacterial Disease, Hygiene Measures.

11. **Animal Welfare** **282—297**

Pit Ponies, Experimental Animals, Laws to Prevent Cruelty, Welfare of Farm Animals, Codes of Practice, Farm Animal Welfare Council, The Law, Increasing Public Concern, Possible Improvements, Multilations, White Veal Calf, Export of Animals, Welfare of Animals being Slaughtered, Conclusion.

Index **298—300**

1

INTRODUCTION

The role of the stockman in maintaining health amongst the animals or birds in his charge, along with his capacity to detect quickly any deviation from normality, is of the greatest importance. As systems of animal production become more intensive, with larger numbers of animals being kept inside wholly dependent upon the person in charge for their feeding, water supply and general comfort, it is clear that the stockman's job becomes ever more crucial in the promotion of high standards of animal welfare and in helping towards the financial success of the enterprise. The need for greater technical knowledge in order to properly control ventilation and heating systems or to administer drugs and feed accurately is clearly necessary. More complicated records need to be kept for greater numbers of animals. Nevertheless, the basic stockman's skills are concerned with the state of well-being of the animal and this remains a mater of accurate observation allied to knowledge of the signs of good health in the particular species with which he is dealing. All these matters must be understood by the veterinarian and any other person who seeks to be of service to the animal industry.

SIGNS OF HEALTH

Horses

A general alertness should be shown so that the animal is aware of what is going on around him; the head should be held high, with the ears making frequent movement backwards and forwards. It should move about freely and there should be a clear bright eye. There must be no sign of saliva or of running eyes or any other discharge. A horse at rest will breathe regularly 8-12 times each minute with no sound as the flanks rise and fall. Horses should have a bright coat and the skin should move easily. A hunter or racehorse will show no excess flesh and the muscles will feel hard. After heavy exercise there will be sweating but this is normally thin and watery, not soapy, with a lot of lather.

The healthy horse shows a keen appetite, grazing frequently if free, but anxious for meal time if stabled, and the dung it casts will be in balls which should break as they fall to the ground, with a non-offensive smell. The urine of a horse is light yellow and between 1-2 litres should be passed without difficulty at a time.

A healthy foal shows great and persistent inquisitiveness, being willing to approach a stranger. It will, however, not allow itself to be handled unless it has already been broken to a head collar. Sick foals are often easy to catch and handle. Again, it is the appearance of the coat which is important and if the foal is well nourished there will be no signs of the underlying bone structure, the animal being rounded in appearance.

The foals of pony breeds kept on moorland or in the hills under traditional conditions will stay with their dams for 6-9 months and continue to suckle if food supply is sufficient to sustain lactation.

Cattle

Cattle are herd animals and therefore, under normal circumstances, will keep together. Any animal that keeps apart should be closely examined. The head should always be held in a normal upright position, any drooping should be looked at with suspicion. When the dairy cow is brought in for milking, it should stand quietly. If there is any nervous twitching of the tail that is not due to fly worry, or 'paddling' of the feet, the animal should be further examined.

The coat of a healthy beast will look glossy or carry a 'bloom', particularly during the summer months when it is likely to be grazing improved pasture, and the skin will move easily over the underlying tissues. The dung of cattle is normally much softer than that of horses and, during grazing in the spring when grasses have a high moisture content, will be very soft indeed. Persistent scouring or diarrhoea from individual animals should always be looked upon with suspicion.

It is normal for cattle to ruminate or to chew the cud. This means that during various periods throughout the day, generally every 4-6 hours, grazing ceases and rumination occurs, when the cud is regurgitated from the paunch and chewed, each bolus taking several minutes before being re-swallowed. If rumination stops, there is something seriously wrong with the animal.

When cattle lie down they first go down on their knees, put their hind legs under them and sink down on to their chest or sternum, sometimes rolling temporarily over on to one side. When they get up they rise on their hind legs first and often stretch before moving off. It is quite normal for cattle to lick their own coats or even others in the herd. Cows in milk must show a consistency in yield from day to day. If there in any sudden drop, except a small one when it comes into season at 21-day intervals, detailed investigation must follow.

In the dairying districts of temperate developed countries, the calf from a dairy cow is taken from its dam within one or two days of birth after it has had the advantage of suckling its dam's colostrum. The practice of removing the calf at birth and giving it colostrum from bucket or bottle is not now so common since it makes much higher demands upon labour. Calves from beef breeds kept under extensive conditions suckle their dams for 6-8 months. Healthy calves have a bright full eye, a glossy coat, are active and show much curiosity when they are up and about, but spend the majority of the day resting. There should be no coughing or scouring. Any sick animal must be removed and housed separately.

Sheep

All sheep should show an alert appearance with bright clear eyes. They will show a good appetite with an obvious tendency for the flock to graze together, except under the special conditions of hill-grazing. Here sheep often graze with some distance between them because of the sparseness of the pasture.

There should be no lameness amongst any in the mob. Foot-rot and other abnormalities of the feet are common in many countries but can be controlled and even eradicated from a property. The fleece must be even. Some of the signs of external parasitism, in the form of lice infestation, fly strike or the more serious sheep scab and scrapie, reveal themselves through intense skin irritation and hence the fleece soon appears uneven. There may be piece of wool above the level of the fleece or, with fly strikes, a darkening of the wool in these areas.

Sheep have a higher temperature (40°C) than cattle or horses and it is normal for them to breathe fast, especially under warm conditions. This is the normal way for them to eliminate excess body heat since, unlike horses and cattle, they do not sweat. The fleece acts as a valuable insulating medium in dry very hot countries.

Lambs born from ewes appropriately fed will quickly be on their feet and suckling. Insufficient food or an unbalanced diet given to ewes during the last two months of their pregnancy is likely to result in small weak lambs being born and in their dams having insufficient milk. A high mortality at lambing time is the result. In health they are active and will spend periods during each day playing, 'gambolling' with others in the group. Any ill-fed or sick lamb will be found on his own, and must be removed and given treatment in separate quarters.

Pigs

Properly fed and housed pigs spend a large proportion of the time resting; but when they are roused they should be active and move about freely. Any staggering or stiffness must be closely investigated. The skin of the pig is a good indication of health. It should be hard without any cracks or scurfiness, spotting or discolouration. If conditions of housing are good with sufficient bedding or at least a dry area for lying if bedding is not provided, the skin will be clean. The snout is normally moist and the ears cool but not cold when felt. The tails of healthy pigs are curled except when they are lying down. The droppings (faeces) of the healthy pig will be formed but soft. A complete balanced diet, based upon ground cereal products, is most commonly fed and this will normally prevent constipation. Any scouring is a danger sign and must be promptly investigated. Affected animals should be segregated at once.

Dogs

A healthy dog shows a keen interest in his food and quickly consumes it if it is of the proper nature. The coat is bright, with some breeds showing a lot of 'bloom'. Persistent scratching of the coat is a clear sign of a high infestation of fleas. Ears must also not be scratched since spreading of infection to the middle ear is not uncommon. There must be no discharge from the eyes and the nose is normally moist. The teeth should be clean, a state that is preserved by seeing that the dog has bones to chew at regular intervals. In very old dogs there may be some discolouration at the base of the canines. The breath must not be offensive. There should be no thickening of the anal muscles of the dog, nor should he drag himself along the ground for these are indications of infection of the anal sacs. The faeces of dogs are normally hard and pale in colour. Temporary diarrhoea may occur following a change of diet, the eating of an excess of food or unsuitable carbohydrate.

Cats

Cats are more active than dogs and a healthy cat will be seen on the move for much of the day. Again, the coat must be clean and bright and if there is much scratching it would indicate an infestation with fleas. The eyes must be clear and have no discharge, and the appetite keen. Normal cat faeces are dark, formed and scft. Any persistent diarrhoea indicates infection; the cause must be diagnosed and speedily treated.

Poultry

Broilers

The remarkable growth of the broiler industry with production being concentrated within large units, means that many thousands of birds are kept indoors, for the most part on a deep litter system, from one-day-old to the time for slaughter at 8 weeks. It is customary to provide light for 23-24 hours each day, the light being either wholly artificial in a controlled environment building, or part-daylight, part-artificial where buildings have windows or, in tropical areas, wire sides. They are thus encouraged to eat a maximum amount of high energy/high protein food to make them grow fast.

Young birds are the most restless of all farm livestock, moving from feed-trough to watering point, scratching in the litter, moving away from the more dominant birds, with only short periods of rest. Although the growth of feathers during this period gives them a ruffled appearance, abnormality is quickly evident if any soiling of the feathers around the vent area occurs. Any unwillingness to move about and a 'droopy' appearance shows that the bird is ill.

The spread of many viral and bacterial diseases can be swift under the warm crowded conditions that are typical of a broiler house. The prompt identification of any unhealthy bird, its removal and examination by a veterinarian, if the cause of its illness is not recognized immediately, must be the accepted routine.

Layers

In the developed countries, nearly all commercial laying hens are kept in battery cages, two or three-tiers high along the length of an environment-controlled house. This means that the roof, and often also the walls are carefully insulated to conserve heat during the winter, and, where they are built in areas with hot summers, to try to keep out the mid-day heat. There is controlled ventilation and mechanically operated feeding, watering and faeces removal. In larger units, an endless moving belt in front of the pen removes the eggs to the packing room as they are laid.

All this cuts down labour costs, but it is most important that all birds be inspected twice daily. Any that are standing at the back of the cage and appear listless, with pale comb or any discharge from eyes or nostrils or evidence of diarrhoea must be removed at once. Especially during the early stages of the pullets being caged, there may be feather-pecking. If this is not dealt with, it can progress to cannibalism, with vent and comb-pecking. If any unaccustomed stress is imposed, such as a change in diet, marked temperature change or repeated disturbance, birds may start or recommence this vice.

If the strain of layers is known to be of nervous disposition, de-beaking of the chicks may have to be carried out.

De-beaking

The aim is to remove the lower or distant third of both upper and lower sections of the beak. It is sometimes carried out within a few days of hatching, using a heated guillotine blade on which there should be a guard to prevent excessive beak removal. For strains of layers known to be prone to develop the vice of feather-pecking, later re-growth of the beak may necessitate repetition of the operation. Care is needed to prevent the removal of an excessive amount of tissue, for not only is this painful, but it leaves the bird unable to apprehend its food in a normal manner. Many producers and scientists believe this mutilation to be avoidable by a combination of selection by poultry geneticists for non-aggressive strains and correct husbandry measures of housing and feeding.

HANDLING

Even when farm animals are kept under the most extensive management systems, such as are many types of ponies, beef cattle or wool sheep, there is need to have them occasionally confined. The manner in which yards, pens and crushes may be constructed. Here it is necessary to consider how each type of farm animal and the common domestic animals should be held or restrained for such purposes as a veterinary examination, to allow a stockperson to carry out permitted minor surgery such as castration and the docking of sheep or baby pigs and to permit the administration of medicines or vaccines.

It is a simple fact that the more skilled and gentle one becomes in approach and subsequent handling, the quicker and easier it is to accomplish one's object. Some species (horse, ruminants, poultry) are likely to be reassured by the sound of quietly spoken words as the initial approach is made, but that approach must be made slowly and deliberately, never in a tentative or abrupt manner. Speed in the subsequent manoeuvres will come with practice.

The overall aim is the employment of the minimum of force to achieve control. If this necessitates keeping the animal immobile, it is important to use a method of restraint that causes the least discomfort. It must also be maintained for the shortest time. Special care is necessary when ruminants have been cast to the ground and immobilized, so that pressure on the large reticulo-rumen or paunch, stretching along most of the left flank, is avoided or minimized.

Horses

If possible one should find out from the horse's usual attendant whether the animal is normally quiet with strangers or if there is any difficulty in catching it. The proper approach is to let the horse see you coming towards it, be sure to hold no stick in your hand and, by speaking quietly, it should be possible to come up close to it on the left or near side. If the horse already has a head collar or halter on,

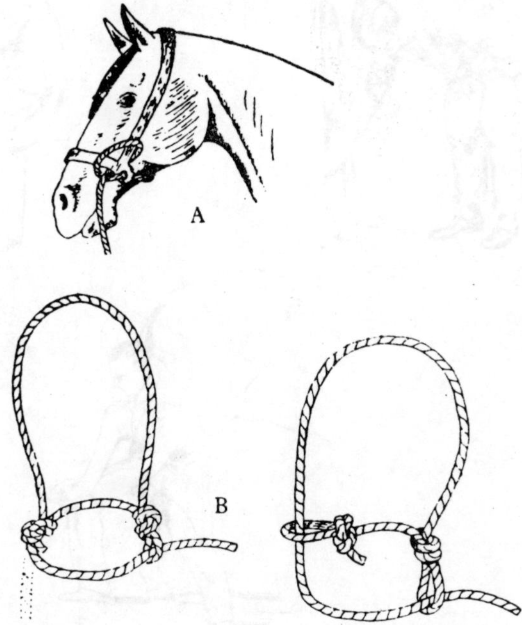

Fig. 1.1. Horse halters. A—Webbing halter, showing cinch to prevent chain piece becoming too tight. B—Two ways of making a rope halter.

then of course there is likely to be no difficulty in catching it, but if not, a halter must be put over its head. This is done by first slipping the rope or shank of the halter round the neck, by holding the forelock, or by resting the right hand on the neck just behind the ears. Then the nose piece is slipped on and, with an extension of the same movement, the poll piece is slipped over its ears. The halter should be prevented from becoming too tight by having a hitch or cinch put in the shank as it leaves the nose band.

If the horse that has to be examined is running with others in a field and, if there is any difficulty in approaching it and putting on a halter, it is probably easier and quicker to drive two or three horses together into a yard and to put the halter or head collar on in the corner of the yard.

The foals of thoroughbreds, hunters and other light horses are normally handled at a very young age and then have a head collar put

Fig. 1.2. Lifting hind leg of horse.

on them so that they are easy to catch. Nevertheless, it is important not to make any sudden movements near a foal, particularly if you want to catch it, since this will only scare it away. If it is frightened a lot by these initial approaches, it is likely to be nervous and suspicious when subsequently being caught.

When a horse is to be examined, it is most important that there be sufficient space at the spot chosen and plenty of light. It is therefore better to examine a horse outside a stable unless there is more room than is to be found in the usual loosebox (3 m × 3 m). The horse's head should be turned towards the door if he is being examined inside. Always remember that any handling of the limbs must be preceded by running the hand down from the horse's head towards the limb that is to be lifted. With a particularly fractious horse it may be necessary not only to have the head held when the hind feet are to be handled, but also to lift a foreleg on the opposite side. Whether fore limb or hind limb is to be examined, one must stand sideways, close to the horse and with one's back to its head.

Cattle

In general it is only dairy or beef cattle being prepared for appearance at Shows that are accustomed to being haltered and led. The most satisfactory way of restraining cattle is to drive them into a crush where the head is held in stocks and the whole animal firmly confined. Without a crush the head has to be controlled by means of a firm grip within the nostrils with the thumb and finger or the use of bull-holders, at the same time holding the horns or the ears of the beast with the other hand or with the help of an assistant. It is of course important that damage is not done to the mucous membrane of the nostril; finger nails should be kept short.

If a halter is to be put on cattle, it is wise to put the poll piece of the halter over the horns and ears before adjusting the loop around the nose, otherwise it may be impossible to get the poll piece over the horns should the beast begin shaking its halter loose when feeling it going over its head.

It is important that some simple form of holding pen or crush be available on dairy and beef properties. Only in this manner can cattle be properly examined and much effort in struggling to hold the head of large cattle avoided. It is undesirable to have to lassoo a beast in an open yard and pull its head round to a post or pillar or through a ring in a wall. This is painful for the animal, it has the incidental

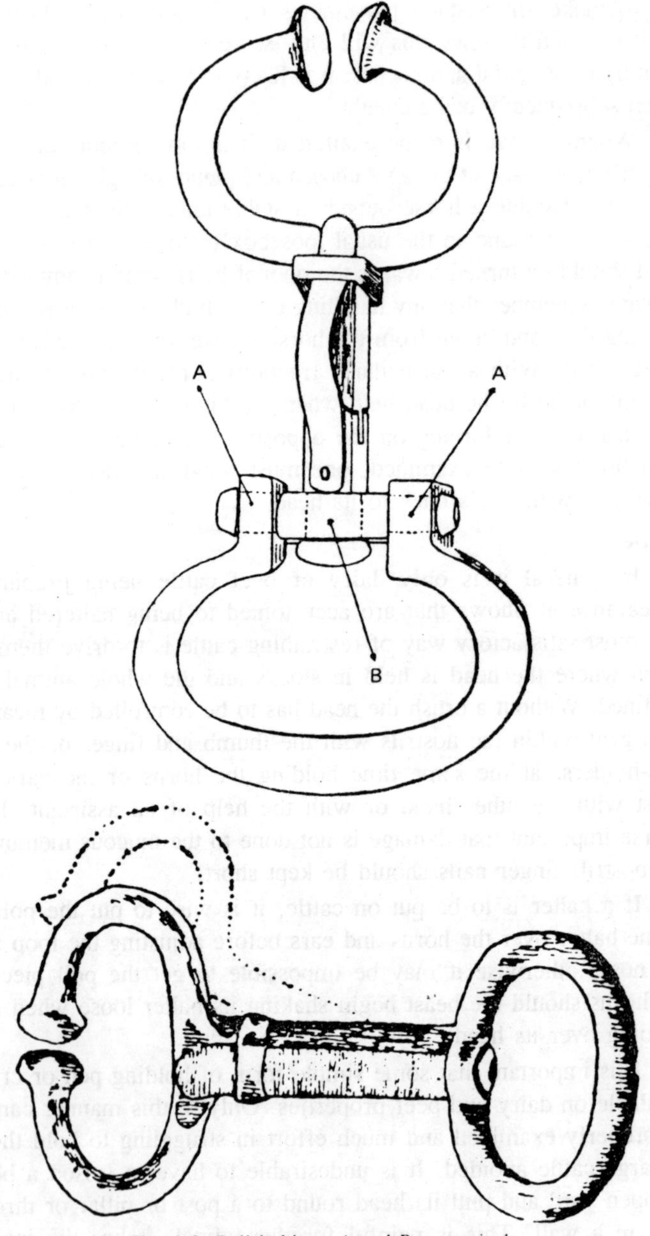

Fig. 1.3. Bull holders. A and B are swivels.

effect of choking it and it demands considerable physical effort on the part of the stockperson. There are many satisfactory designs of cattle-crushes.

Nose rings

Bulls

It is important that all bulls being kept for service have rings placed in the cartilaginous septum of the nose before they are one year old. The common type consists of two semi-circles hinged at one end and generally made of copper. The free ends dovetail into each other and are secured either by a flush spring or a screw with a countersunk head, so that the inserted ring is entirely smooth. A hole must first be punched with a special instrument at least 1.5 cm from

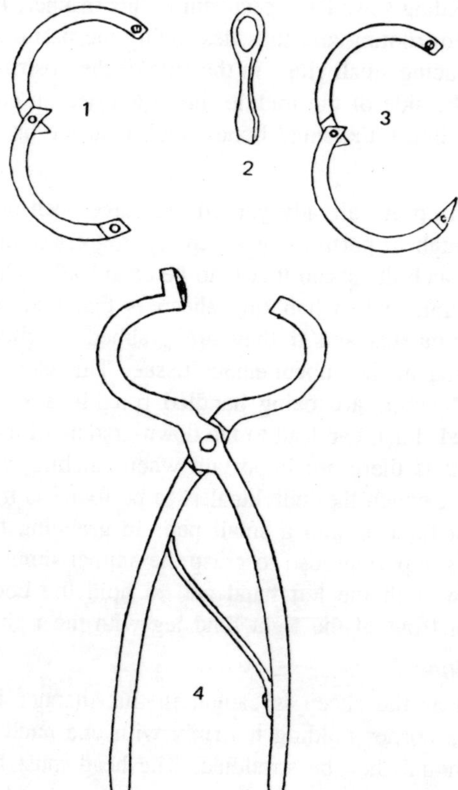

Fig. 1.4. Bull rings and nose punch. 1. Standard bull ring. 2. Key for screw fastening. 3. Self-piercing bull ring. 4. Nose punch.

the nostrils. It is possible to avoid having to use a punch by the use of a ring that has one end made into a sharp point. This is thrust through the cartilage of the internasal septum and then closed and secured in the usual manner. It is important to examine the bull ring periodically since they wear and occasionally can be partly torn from the cartilage. No bull should be handled unless the ring can be seen to be firmly attached and unworn. A bull pole having a controllable clasp at its end can be used to get hold of the nose ring and bring the bull into a pen or crush. Bulls must always be treated with great caution since their behaviour is uncertain.

Pigs

Very few pigs are now kept under outdoor systems but, in the past when breeding sows have been run in this manner, it was desirable to prevent them rooting and thus destroying the pastures. This can be effected by placing small rings in the side of the nostrils. They should be put in at the side of the middle line often one on each side, being careful not to injure the small bone which is at the base of the snout.

Sheep

Unless sheep are already yarded and have been put into a 'race' just wide enough to permit one sheep to progress along it, it will be necessary to catch the sheep that is to be examined. What must always be borne in mind when handling sheep is that their skin is thinner than other ruminants and if they are grabbed roughly there can be serious bruising of the subcutaneous tissue. This can be of economic importance if lambs are being handled roughly a short time before being marketed, for it can lead to the down-grading of a carcass because of bruising. It is therefore important when catching a sheep that the group amongst which the individual is to be found is manoeuvred into a corner of a field or into a small pen. In grabbing the sheep to be examined it is important also to grasp the animal simultaneously under the lower jaw with the left hand and to hold the body in the flank region just in front of the right hind leg with the right hand.

Holding position 1

As soon as the sheep is caught in this manner it is as well to back it into a corner holding it firmly with one hand under the jaw. The sheep should then be straddled. The head must be controlled at all times. Avoid grasping any sheep, particularly lambs, by the wool. If there has been at any time an interruption to its food intake, there will be a weakness or fault in the fibre, so the wool will break easily at this point. With lambs there will be bruising.

In this position one can examine the teeth, or the state of the visible mucous membranes around the eye to indicate whether there is any anaemia or jaundice. The normal colour is a bright pink. In this position sheep may also be drenched, or a blood sample taken from the jugular vein.

Holding position 2

If the feet of a sheep are to be treated for foot-rot, if the udder of the ewe is to be examined or the health of the ram's testes checked, it is necessary to 'turn up' the sheep, i.e. it is to be sat on its haunches. This is done by holding one hand under the jaw and, standing on the left side of the sheep, with the right hand grasping the right flank, the animal is momentarily lifted off the ground and brought on to its rump.

A second method of achieving this position, suitable for use with large rams or a small operator, is to stand on the left side of the sheep with the left hand holding the animal's head and, with the right hand, reach underneath the right flank and grasp the left hind leg just above the hoof. By pulling on this leg and simultaneously stepping backwards the sheep can be turned up without difficulty. Do not allow the sheep's head to fall back unsupported through the legs of the person holding the sheep. This frequently provokes the sheep to struggle.

Docking and castration of lambs

Most lambs have a part of the tail removed and male lambs that are to be marketed are still castrated. The lambs of hill breeds are often not docked. Both operations may be accomplished by placing rubber rings over the scrotum and the tail by means of an applicator. Use of this method, although bloodless, causes a fairly prolonged period of discomfort, and is limited, by law, to performance during the first week of the lamb's life. Docking may be carried out a few weeks later by means of a knife and castration by either a knife or emasculator.

The lamb is lifted off its feet by being held under the brisket and the hindquarters placed on a board attached to a convenient fence. The lamb's right foreleg and hind leg are held by the right hand and the lamb's left foreleg and hind leg by the assistant's left hand, the legs crossing over and being held by the shank or cannon bones.

If a knife is to be used, it must be very sharp and made free from pathogenic micro-organisms by boiling in water in a container with a tight-fitting lid for at least 10 minutes. Between each lamb, the knife should be placed in a disinfectant solution, renewed from time to time.

The tip of the scrotum or purse is grasped and quickly removed. This permits the two testes to be forced out, gripped at their base and their connection with the spermatic cord broken with a sharp twist upwards.

In docking the tail, it is desirable to make the cut between the third and fourth coccygeal vertebrae. This avoids passing through bone and leaves sufficient tail to cover the vulva of a ewe lamb. The skin of the tail should be pushed upwards towards the base of the tail before the cut is made, for this helps to form a flap of skin to cover the end.

If the bloodless castrator or Burdizzo is used, it must be applied carefully to the neck of the scrotum, feeling with the free hand to be sure that the spermatic cord on that side will be within the jaws of the instrument before closing it. The same procedure is repeated on the other side. If care is not taken with each lamb, misses will occur, with subsequent need to repeat the operation later, possibly at the stage when an anaesthetic is necessary.

Welfare legislation varies from country to country, but, in the U.K., it is only possible to use the rubber ring method for castrating and docking during the lamb's first week; the above methods can be used without an anaesthetic up to the age of three months.

In many tropical countries, these operations are not undertaken partly perhaps because of the fear of infection, but mainly because there is no butcher's prejudice to the carcass of the ram lamb. Drier forage feeding prevents formation of the soft faeces that follow grazing in wetter climates with the consequent faecal build-up ('dags') on the wool of the long tail.

Pigs

Nearly all pigs are now kept indoors all the time, so that it is not too difficult to take a hurdle into the pen and to segregate a pig in a corner. A small loop or rope attached to the end of a short stick, similar to the twitch used with horses, can then be slipped into the pig's mouth and tightened just behind the upper tusks. Another type of pig catcher is made from an iron bar about 90-cm long having a handle at one end and at the other a 10-cm ring bent round at a moderate angle. The ring is slipped round the upper jaw behind the tusks in the same position as the rope noose.

Small pigs that have to be caught for examination or for castration may be grasped by the hind leg above the hock or even by the ears. For castration an assistant may hold the hind legs of the pig in each hand, restraining the forelegs crossed alongside the hind legs. The pig can then be sat down on a board or it can be held in his lap.

Dogs

It is always desirable that the dog's owner holds the animal for any examination that needs to be made. Even so, it is usually necessary with nervous dogs or if any injury or infected areas have to be examined, to secure the dog's jaws with a tape muzzle. This is simply made from a piece of bandage about 1-m long on which two half hitches are created in the middle. These are slipped over the dog's nose from behind and pulled sufficiently tight to prevent the jaws opening and the ends tied behind the dog's ears. If the head or front of the dog is to be examined, the animal can be placed on a table and held gently by an assistant from behind steadying the head with one hand and holding the forelegs above the knee with the other. If the abdomen is to be examined, the assistant holds the right and left forelegs above the knee with the left hand from in front and the two hind legs with the right hand reaching over the dog's back to do so. The dog may then be rolled over on to its side or on to its back with the legs extended if this is necessary.

If a vicious dog has to be brought under control, a dog-catching device with a piece of rope tied at one end to a long pole and, a little behind this, a ring through which the rope can pass forms a loop and this can be put over the dog's head and drawn tight so that it can be brought close enough for a muzzle to be put on.

With efficient tranquillizing and short-acting anaesthetics being readily available, any prolonged examination should be carried out after the dog is sedated.

Cats

In most instances, it is desirable to restrain a cat to the smallest extent and to take time to calm it by fondling the animal. The mouth can be examined fairly readily by holding the animal gently but firmly with the left hand from behind, holding the head steady and pressing gently on the two sides of the jaw; the right hand can then gently press down the lower jaw to gain a rapid but generally adequate look inside the mouth. In the same way, any medicine or tablets can be quickly inserted without causing distress. For castration or the undertaking of more painful examinations, it is nowadays most convenient and humane to use a quick-acting anaesthetic injected intraperitoneally.

Poultry

The ease with which one can pick up and examine a fowl, whether broiler, grower or layer depends upon the strain of bird. Generally speaking, it is those strains in which the heavier breeds have been incorporated that will, on being closely approached, crouch down and stay motionless, so allowing one to pick up the bird.

With more nervous strains (the lighter breeds), to pick out an individual from a constantly shifting group will require the use of a length of stout wire, at one end of which is a narrow loop or long hook; alternatively a shorter piece of wire may be attached to a long pole. The hook is slipped on to the shank of the bird being sought and drawn forward.

If it is necessary to have a free hand in order to examine the bird, this can be done by placing a hand under the base of each wing and holding the bird in the crook of the arm, pressing the bird against the examiner's side. It is to be remembered that older cockerels, especially broiler breeders, may have sharp-pointed spurs, making this manoeuvre somewhat more hazardous. Use of leather gauntlets is advisable.

Restraining Animals in Buildings

Headstall or Head Collars

This is a method of securing a horse in a stable or stall. Today, for cheapness, ease of cleaning and durability, coloured nylon head collars are increasingly being used.

Neck straps or neck ropes are often used in stables but, whether it is a halter or a neck tie, the shank in each case is put through a ring or hole in the manger, with the end attached to a metal or wooden 'sinker' to prevent the horse from putting his foot on the head rope, yet allowing the maximum length of movement.

Horses may be tied to the rings that are often to be found in stalls or on the heel post. When tying a horse in a stable it is best to use a knot that can be released quickly. It is done by doubling the head rope or the shank of the halter with one short end. This doubled piece of rope is put under the ring and then another loop in the long end nearest the horse is inserted through the double end and the short end is pulled tight. A second loop is made in the short end and this is

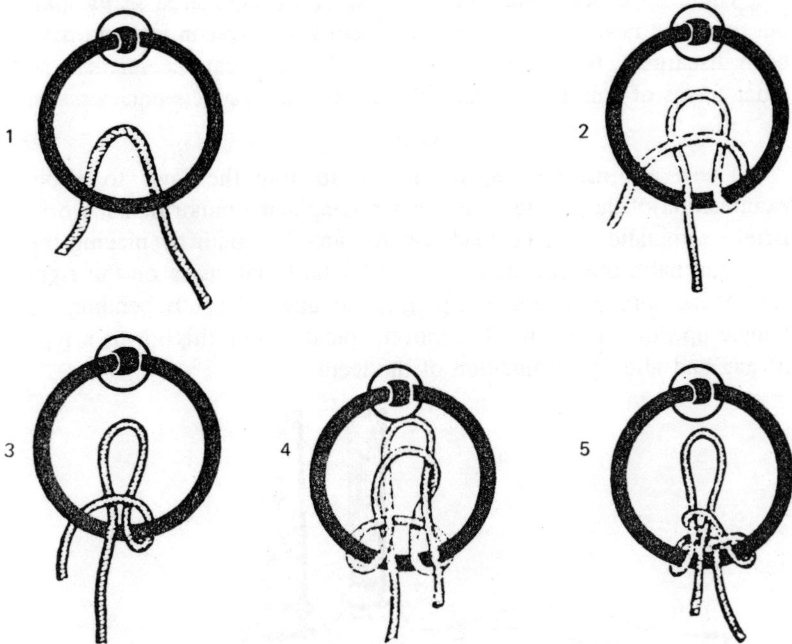

Fig. 1.5. Method of tying horse lead to ring.

put through the first loop and tightened by a pull on the long end. A quick tug on the short end releases the horse.

Twitches

The purpose of putting a twitch on a horse is to cause a sharp focus of discomfort in order to divert the animal's attention while some necessary manoeuvre is undertaken in the horse's interest that may be momentarily painful. Occasionally it may be used if the horse is particularly restive and hard to handle. A twitch must only be used very sparingly and be kept on for the minimum of time.

It consists of a short handle about 1 m in length in the top of which is a hole bored about 5 cm from the end. Through this hole is a short piece of rope about 45 cm in length. The horse must be held when the twitch is to be applied and it is better to have it backed into a corner of a loosebox or a stable yard. The loop of the twitch is placed on the spread-out fingers of the left hand which then grasp the upper lip of the horse without involving the nostrils. The right hand rapidly twists the stick to tighten the loop which now grips the lip. The stick should point backwards towards the horse's left shoulder.

There have been many other forms of twitches used in the past but these are nearly all designed to cause greater discomfort and have been discarded. Today the greater availability of local anaesthetics and other forms of pain relief makes the use of such gadgets unnecessary.

Gags

These are mechanical means of opening the jaws to allow examination of the mouth or for undertaking some minor dental work. Brief examination can be made of the horse's mouth by placing the left hand palm downwards through the interdental space on the right side of the horse's mouth, grasping the tongue and gently bending the tongue upwards as the hand is moved round so that this forms a type of gag and allows examination of the teeth.

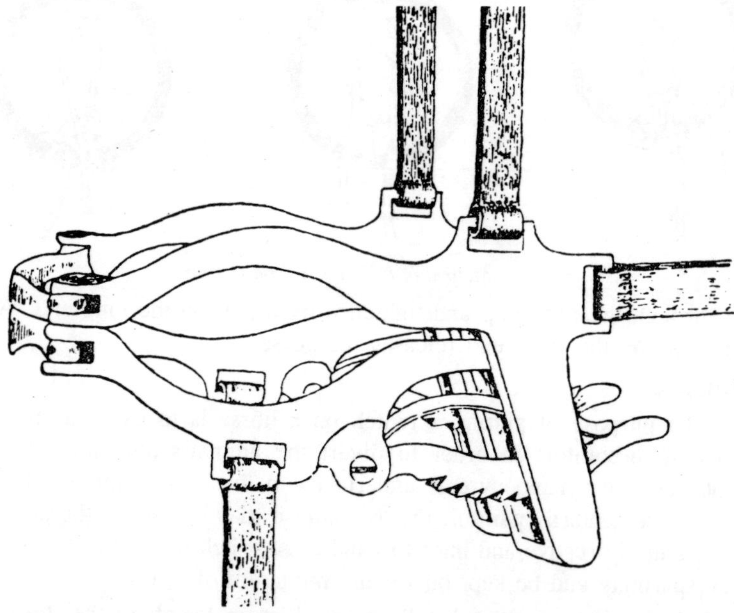

Fig. 1.6. Hausemann's horse gag.

Probably the commonest type is the American gag or Hausemann's gag. The two short platforms rest on the tables of the upper and lower incisors with these hinged and provided with a ratchet. It easily comes to pieces for transportation.

Cattle

It is very useful to be able to open the mouths of cattle and to remove any food such as pieces of root crops which may have lodged

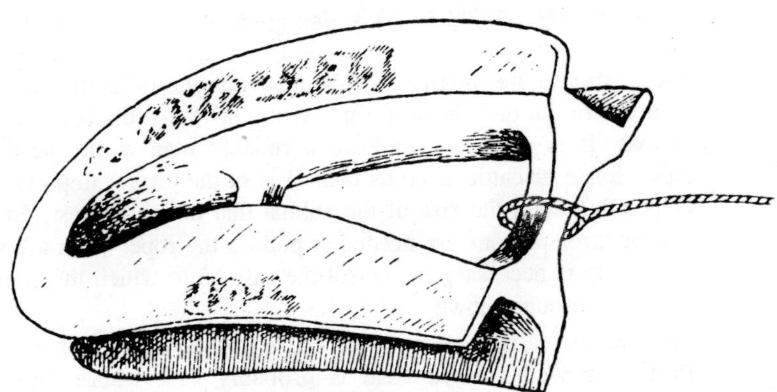

Fig. 1.7. Cattle gag.

in the larynx. Drinkwater's gags are the commonest type available. They are made of aluminium, one for each jaw. They correspond to the shape of the jaw and are fitted with flanges to accommodate the upper and lower cheek teeth. When being inserted the right gag is held in the left hand and vice versa. It is desirable to have a cord attached to the front part to help in the gag's removal.

Dogs

Spring gags are rubber-covered and fit into the side of the dog's mouth and when released force the mouth open. They are meant to be placed upon the points of the upper and lower canines on one side of the mouth. There are other types of gags, one being a smaller modification of the common horse gag.

Casting Animals

At one time it was customary when operations had to be performed on the feet of horses and cattle or for other purposes, to first cast the animal. Nowadays the almost universal custom is to use a combination of tranquilizers and anaesthetic agents with the animal in the standing position. This means that, at the appropriate moment as the anaesthetic takes effect, the animal can be placed on an operating table in a surgery, or if an operation is to be carried out on the owner's property, the animal is placed on a prepared bed. However, under exceptional conditions, it may be necessary to cast a conscious animal and one or other of the following methods may be used.

Precautions before Casting

1. If possible, it is desirable to fast an animal for up to 12 hours. It is clearly undesirable to have the stomach full of food at this time.
2. There should be sufficient labour available to complete the manoeuvre in the shortest time with the least distress to the animal. It is necessary to have a reliable man at the head of either horse or cattle since its control is of the utmost importance. Depending upon the size of the animal that has to be cast, either two or three persons are needed to pull on the rope. With a horse extra help is necessary to control the side on to which the animal has to be brought down.
3. The site for casting needs to be chosen as carefully as possible. Part of a well-grassed field is probably best where there is sufficient space and no objects against which the animal could cause itself injury. In bad weather or under other circumstances, a bay of a barn or a large loosebox can be suitable. There should be plenty of clean straw put down so as to make the animal's fall as comfortable as possible. In veterinary hospitals a casting bed of peat moss, tan bark or sand is often available. The space needed to cast cattle is much less than for a horse.
4. Stable bandages are advisable for horses, coming well down the legs and being properly secured. Sometimes kneecaps and hock boots may be put on as well. Tail bandages can be useful with hunters or racehorses to prevent the tail getting caught in any ropes and of course any operations in the perineal region necessitate the tail being bandaged.
5. A strong webbing halter or head collar must be used. No bit can be in the horse's mouth and any twitch that has been used in adjusting ropes must be removed.
6. Any operations on the feet necessitate the removal of shoes. This is particularly necessary if hobbles are to be used in casting.

Hobbles

This is the method still used with heavy draught horses. A set of hobbles consists of four with a chain about 3-m long to which is attached a rope of about 5 m and a spring 'D' catch. One of the hobbles, 'the master hobble', has an arrangement by which the chain is screwed on to it.

The hobbles should first be laid out, each in its right position, and then placed on the horse. If the horse is to be cast on the right side, the master hobble is attached to the left fore pastern since this will be uppermost.

A rope is slipped round the upper arm of the foreleg which is to be uppermost and passed over the withers to the assistant standing on the other side whose task is to pull the animal on to the required side. As soon as the hobbles are adjusted, the man at the head backs the horse quickly; at the same time the hobbles are pulled together and the man holding the rope over the withers pulls the horse over.

As soon as the operation is finished, the hobbles must be taken off with the horse still lying. Only after they are all removed can the man at the horse's head release this and the horse will, if not still suffering from the anaesthetic, rise.

Casting by sidelines

This is the method used with unbroken and light horses especially if they are nervous and likely to resent hobbling. A rope about 13-m long is laid out on the ground and doubled. A loop large enough to go over the horse's head is made in the centre with a 'figure-of-eight' knot, designed to lie flat against the chest of the horse. For most

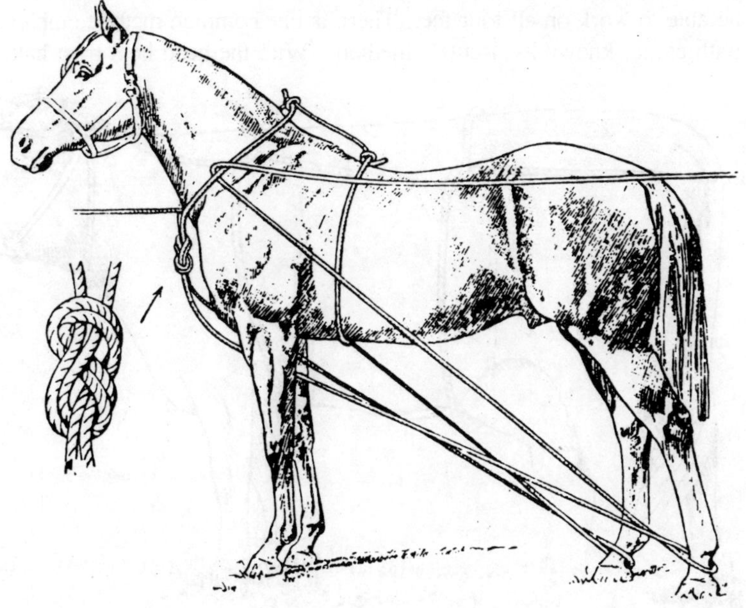

Fig. 1.8. Sidelines for casting horse.

yearling colts the loop should be about 45-cm long to make a 1-m circumference loop.

The loop is slipped over the horse's head and the two coiled ends of the rope are passed between the forelegs and round the hind legs to rest above the hocks. This clearly requires the aid of two assistants. The ends are then brought forward underneath the first part of the rope and through the neck loop. One rope is now held by two persons in advance of the horse and the other one by two others well behind them. The ropes are slipped down to the pasterns and as the order to pull is given the man at the head backs the horse as the hind feet are pulled forward, forcing the animal into a sitting position from which he can be turned over on the required side. The man at the head extends it immediately the horse is on the ground. The upper hind leg is drawn right up to the shoulder and two or three half hitches are taken round the pastern close to the hind leg. The animal is then turned over and the other hind and forelegs similarly secured.

Casting cattle

Cattle are also not often cast nowadays, but occasionally when operations have to be carried out on the feet of cattle it is helpful to be able to work on all four feet. There is one common method employed with cattle, known as 'Reuff's method'. With the head held by a halter,

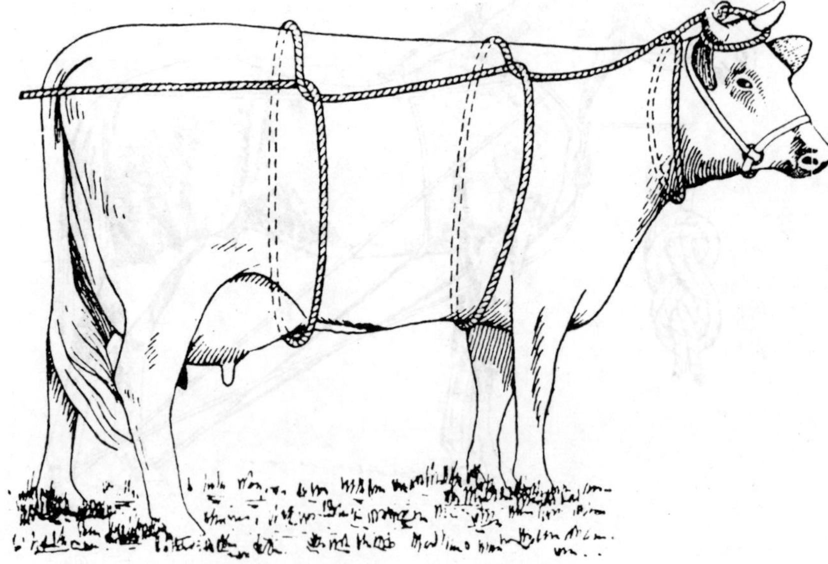

Fig. 1.9. Reuff's method of casting cattle.

a rope of about 10 m is taken and, if the animal has horns, a running noose is passed around their base. With most cattle now hornless, the noose should be eliminated and a loop of the rope placed round the neck and secured above the withers. The long end of the rope is taken along the back and a half hitch made around the chest immediately behind the front legs and another half-hitch around the abdomen in front of the udder or scrotum.

The rope is pulled by one or two men and the beast will normally collapse on to the ground. The feet may then be secured to any fixed object. Again, it is necessary for the head to be kept extended and under control.

Administration of Medicine or Drenching

Horses and Cattle

The administration of anthelmintics or medicine to horses or cattle can generally be achieved by a quiet approach in familiar surroundings with the head under control. This will mean having either a head-stall or a halter on a horse, and with cattle that the head is held in a milking-bail or crush.

It is desirable to have the drench or medicine in a plastic bottle with a fairly narrow neck. Some worm drenches for horses are put up in paste form in a plastic syringe. If many calves or cattle have to be given an anthelmintic, a drenching gun will be used. Standing just to the left of the horse and to the right with cattle, the bottle is placed in the mouth above the tongue entering at the point of the interdental space just behind the incisor teeth. Placing the thumb of one hand inside the opposite interdental space will have the effect of encouraging the animal to open its mouth sufficiently to insert the neck of the bottle. Exercising care and patience, the animal will generally swallow without difficulty. It is wise only to give as much at one moment as can be conveniently swallowed and to pause before giving the next dose. With horses, when a lot of fluid has to be given, it is better to pass a stomach tube and slowly administer the medicine in this manner. A semi-stiff tube either of plastic or rubber whose external diameter should not be more than 1 cm for foals, 2 cm for ponies and 3 cm for horses. It is lubricated and then slowly passed along the inner edge of either nostril. As it enters the pharynx the animal will normally make a swallowing movement and this will assist its passage into the oesophagus. Before any fluid is administered one must be sure that it has not entered the trachea. The sound and feel of the movement of air will be detected if this has occurred. The tube must then be withdrawn and re-inserted.

Pigs

This is an uncommon need in pigs since most anthelmintics can be given in the feed. But if a medicament is to be given by mouth, it may be necessary for an assistant to put a rope noose in the pig's mouth behind the upper canines and to pull the head upwards. Now you may straddle the animal and, steadying the head by one hand on its ears, the medicine is placed in the mouth in small amounts.

Sheep

The only common need to drench sheep is to administer an anthelmintic and this is most conveniently done by means of a drenching gun with a mechanism for adjusting to the required dose. The nozzle must be placed in the mouth through the interdental space and over the tongue to a position in the anterior part of the pharynx. Gentle pressure is placed on the trigger mechanism to inject the dose and to wait until the animal has swallowed, if further doses have to be given. If any other medicament in liquid form has to be given, a similar method as described above is employed.

Dogs

Only with fractious patients will it be necessary to apply a tape far down the nose. With a quiet approach it will generally be possible to pour medicine through a narrow bottle into a dog's mouth, but if necessary a forefinger can make a pouch on one side by drawing out the cheek and an assistant can pour the medicine into this to allow it to trickle back to the pharynx.

Cats

If the scruff of the neck is held up as high as possible with one hand and some pressure exerted, the cat will open its mouth and the medicine can be put in slowly.

Birds

A fountain pen filler can be a useful method of administering medicine to birds.

INJECTIONS

Injection has increasingly become the method preferred for administering medicaments and substances used in preventive medicine, particularly all manner of vaccines. Various routes may be employed, depending upon the speed of action and prolongation of effect desired.

In descending order of rapidity of action, the intravenous route is most rapid, followed by the intramuscular and subcutaneous routes.

Subcutaneous Injections

A convenient site for small animals, including sheep, is in the region just behind the shoulder. The skin is swabbed with methylated spirit, a pinch of skin taken in the left hand and the hypodermic needle inserted with the right hand. In cattle and horses, a preferred site is at the base of the neck, just in front of the shoulder.

Intramuscular Injections

A site is selected where there is a large muscle mass, with most species the buttock muscles being suitable. The area is swabbed with a skin disinfectant, the needle detached from the syringe, and is quickly plunged into the muscle mass. With the animal motionless, the dose is injected.

Intraperitoneal Injections

Occasionally this route may be chosen for the injection of large quantities of suitably dilute solutions. After skin disinfection an area in the upper flank is generally used for large animals, but for small animals which are held on their side or back, a suitable spot on the abdomen may be selected.

Intravenous Injections

In horses, cattle and sheep the jugular vein is commonly chosen for the injection of drugs at appropriate concentrations when an immediate effect is required.

A point mid-way along the neck is prepared by clipping, necessary with sheep and cattle and horses with winter coats, and swabbed with spirit. The vein can easily be extended in horses and sheep by placing the thumb in the jugular furrow a short distance posterior to the prepared site. With cattle, especially with bulls or the heavier beef breeds, it is necessary to place a cord with a noose around the base of the neck and to tighten this until the vein distends. Once the vein is clearly visible, a sharp 18 g × 5 cm needle, bevel upwards, is inserted into the vein, the syringe attached and the injection made. The distending pressure is simultaneously removed. The injection completed, the thumb of the left hand is pressed over the point of entry to the vein as the needle is quickly withdrawn. There will normally be no bleeding.

Occasionally, if a cow is recumbent, it is convenient to inject into the mammary or coccygeal veins. In the dog and cat, the radial vein in the forearm or the external saphenous vein of the hind leg may conveniently be chosen. A narrow gauge and shorter needle will be used than in large animals.

In pigs, it is inconvenient as it is difficult to make intravenous injections. An ear vein must be located, the area cleaned in the usual manner and a narrow gauge needle inserted. Medicaments for pigs are normally given by mouth and, where a group are to be treated, by mixing with the feed.

Intramammary Injection

Infection of the udder of the cow is an all too common occurrence and a commonly used form of treatment is to place either a single antibiotic or a combination of antibiotics into the teat canal. In many herds it is now routine to treat all cows as soon as lactation ceases and their dry period begins, but in acute cases of mastitis, treatment must be given during lactation.

In such cases the affected quarter must be thoroughly milked out before the medicament is injected. In every instance, all dirt and foreign material of any sort must be washed from the udder, the quarter dried and then the orifice of the teat canal swabbed with spirit. A short cannula is attached to the tube of antibiotic and this is gently pushed into the teat canal and the contents squeezed out. After withdrawal, it is customary to massage the treated quarter for a few minutes to help spread the injected material throughout the gland. It must be remembered that it is a legal requirement in the United Kingdom that milk secreted in any treated gland must be discarded for three days after treatment.

Intra-Uterine Injections

Following infection of the uterus, known as a metritis, irrigation with an antiseptic solution is sometimes prescribed. This requires the careful insertion of a special catheter through the opening from the uterus into the vagina, the os uterus, and the solution run in under gravity. It is normally a two-way catheter so that fluid can drain out simultaneously as it is being flushed in. Great care has to be taken in inserting a catheter into the uterus and the opening must be patent to allow the catheter to enter.

Intra-Vaginal Tampons or Pessaries

The commonest use for medication by insertion of sponges into the vagina is to permit the slow absorption of progestagens, artificial analogues of progesterone, a hormone secreted by the corpus luteum or yellow body of the ovary. The action of progestagens is to prevent the manifestation of oestrus. So for, this has been found to be most reliable with ewes and other means of administering progestagens to larger animals have had to be devised. It is desirable to use a small speculum to dilate the vagina of the ewe during the autumn mating season and the tampon can then be slipped into the vagina. Attached to it is a small tape to aid its withdrawal at the end of a fortnight.

2

HOUSING AND YARDING

The need to provide confinement and shelter of some kind for all types of domesticated livestock is plain. Even in the most extensive systems of husbandry, stock have to be put into enclosures or yards for such purposes as their protection from predators, or in order to separate offspring from their dams before puberty, to run-off animals that must be marketed, to arrange appropriate groups for mating, or to control disease by administering drenches, by vaccination or by dipping.

Under farming systems of a more advanced and intensive nature, well illustrated in pig and poultry production, the genetic potential possessed by modern breeds, strains and crosses can only be realized when the various factors making up the total environment, i.e. the feeding regime, provision of water, space, temperature and humidity, are under the farmers control.

The dairy farmer, even in countries enjoying an equable, temperate climate, such as we find, for example, in New Zealand, must have yards, milking bail and a milk storage room. But in many part of temperate countries and in the tropics, it is necessary to provide housing for cattle and sheep, as well as for pigs and poultry, for protection from winter weather for at least part of the year, or from the sun in the dry tropics, and from heat and rain in the wet tropics.

A further reason for housing fattening cattle and pregnant ewes during the winter, where rainfall is high and soils heavy, is that treading or 'poaching' of the wet ground destroys a proportion of the pasture or, at the least retards and reduces the quantity of grazing available in the spring. The comfort and working capacity of the stockman is greater when protection is provided from unpleasant climatic conditions.

Size

The type of housing and the extent of the handling facility must depend upon the species of livestock, the location and the number of animals or birds involved. The requirements of a small family dairy farm in a European country are satisfied fairly easily. The problems of design are greater when accommodation has to be built for dairy units containing hundreds of milkers or as many breeding sows; while the hundred-thousand units for layers and broilers pose their own problems.

The smaller the number of livestock confined in one place, the less the danger of widespread infection from any pathogen that becomes established. Many attempts have been made to discover if there is a critical number for any species and although some veterinarians are prepared to specify a number, the truth of the matter is that such important factors as the number and competence of the stockmen looking after the animals or birds, the character of the building in which they are housed, the level and quality of the diet and disease control, make it unwise to state numbers.

There are certain practical matters that have to be considered when plans are being made for the housing of various farm livestock and these will be discussed as the requirements of each species is considered in greater detail.

Requirements for Housing

Roof

Since it is shelter from very low or very high temperatures that are the main purposes of putting stock into shelter, it must be understood where the greatest heat loss or heat gain occur within a building. Where the need is to provide protection from the sun, it is similarly important that the insulating properties of roofing materials are carefully

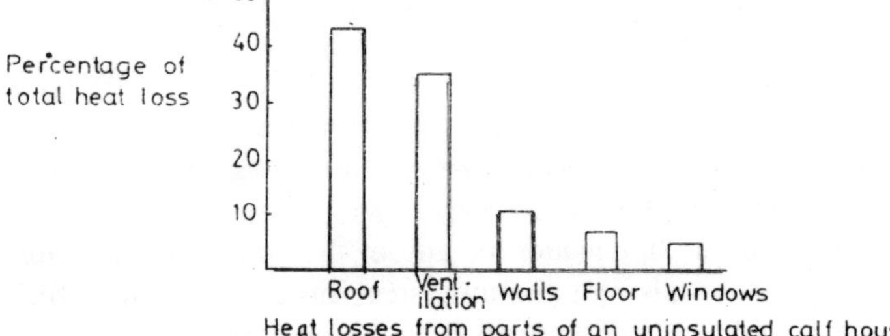

Fig. 2.1. The amount of heat loss from different parts of an uninsulated calf house.

considered. Insulation is measured in terms of thermal conductivity of various materials that may be used in building construction. This is spoken of as the *K* value of the material. It is the amount of heat in watts transmitted through 1 m^2 of the material when the temperature difference of 1 °C is maintained between opposite surfaces of 1 m thickness.

Of more practical use in the measure of the rate of heat transfer through a structure of measurable thickness. This in actual use is mostly composed of layers of different materials having different *K* values. The *U* value is defined as the amount of heat (in watts) transmitted through 1 m^2 of the construction from the air inside to the air outside when there is 1 °C difference in temperature between the inside and the outside. In other words, it is calculated from the *K* values of the materials used and their thickness. The manufacturers normally provide the *U* values of the materials they sell.

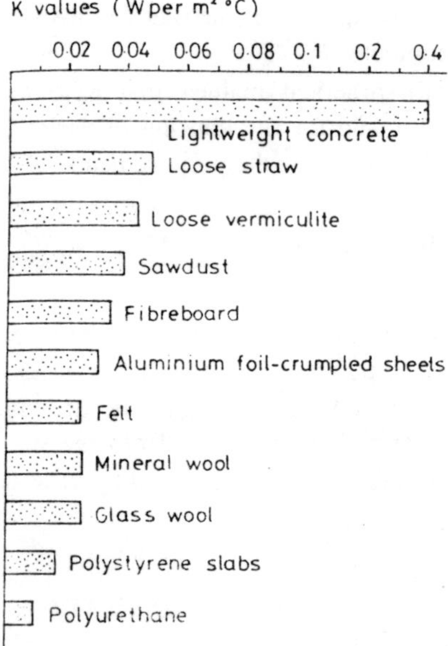

Fig. 2.2. The amount of heat (K value) transmitted through some common materials used in animal houses.

It is clear that a combination of any of the commonly available weather-proof materials, aluminium, steel or asbestos, to which polystyrene or polyurethane is applied to the underside in an appropriately thick layer, held in place by a thin layer of material to form a seal, makes an efficient insulation against heat or cold.

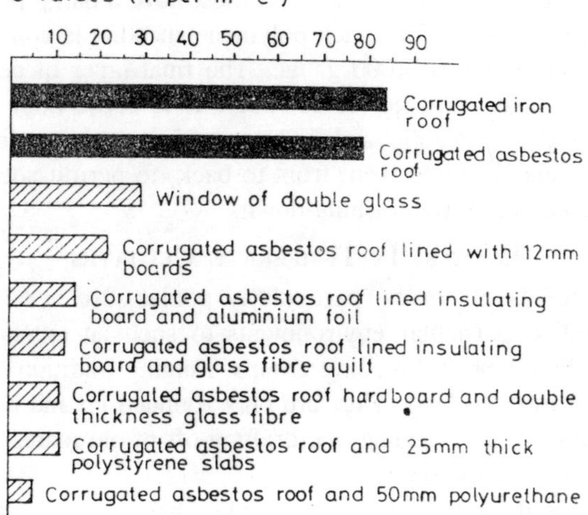

Fig. 2.3. The amount of heat (U value) transmitted through various types of roof used in animal houses.

In many places, there is the additional problem of the effect of condensation or induced humidity from the breath and urine of stock within the building. There is a direct relation between the level of humidity, the difference in temperature between the inside and the outside and the *U* value necessary in the material chosen for ceiling or walls to avoid condensation. A vapour seal should be added between the impervious surface on the underside of the roof and the insulating material. A wet insulating material has no insulating value.

Walls

Walls are not so important in preventing heat or cold gain, but care must be taken:

1. That they are constructed in such a manner that they are able to withstand the physical pressures that stock housed within the building may exert from time to time,
2. They have an impervious and durable surface that can be repeatedly washed down and occasionally be disinfected,
3. That there is sufficient and appropriate ventilation as an essential part of the construction. A dampcourse, usually a bituminous layer, must be inserted in all walls.

Floors

It is important that floors for pig and poultry houses, where the stock are always close to the floor, have an efficient insulating layer as part of their construction. All floors must have a damp-proof course to prevent rising damp, for which polythene sheeting is convenient and effective, usually 500 or 1000 gauge. The final layer of cement over the insulating layer must not be so rough as to cause hoof damage to pigs, but must be non-slip and have a sufficient slope, in pig pens between 130 and 170 mm from front to back, to permit adequate run-off of urine and effective washing-down.

Horses: In Temperate Climates

Some breeds of horses, particularly ponies, are very hardy and well adapted to particular environments to such an extent that they spend the whole year in breeding groups, ranging traditional moorland or preserved forested land. They are only rounded up and brought into yards periodically for branding, the drafting-off of old mares or stallions or for veterinary purposes.

Today, at the other end of the scale, there are very few of the heavy or large draught horses on farms. In the temperate zones, where they were once so numerous and important for the cultivation of crops and for haulage on the roads, there are few to be found following the old routine of work by day and going into stables for the night.

In the U.K., there has for many years been a thriving trade in horsemeat, with both carcasses and live animals being exported to continental countries. Some use is being made of heavy horse breeds for crossing upon light mares with the horsemeat trade in mind. Horses growing and fattening for this trade are commonly kept outside all the time, having to depend upon pasture for most of the year.

It is with light horses used for riding and with Thoroughbreds kept for racing that there is need for stabling. There are still many old and not very suitable or maintained stables being used. Whether the keeping of riding horses and ponies is for private pleasure or as a commercial undertaking for hiring-out, it is important that old buildings be maintained or altered to meet the minimum demands needed to provide the animals with comfortable quarters. Any new buildings clearly must meet these requirements.

The United Kingdom Riding Establishments Act (1973, 1977) imposes a duty upon local authorities to have a veterinarian inspect the premises annually if a licence is to be renewed.

Horses are either kept in (1) *loose boxes*, separate accommodation for each horse, in which the animal may move about freely, or (2) *stalls*, each a division within a stable block of varying capacity, where each horse is tied. This latter was the usual method of housing draught horses. Riding horses, such as hunters, may be kept in this manner, but it is more usual to have them in loose boxes. Racehorses and stallions are always kept in loose boxes.

Loose Box

The usual measurements are 3.5 × 3.5 × 3 m high. If only ponies are to be housed, 3 × 3 × 3 m is sufficient. In temperate climates the roof and wall insulation should have a *U* value of 0.30. The floor must be carefully laid in concrete to give a small slope of 5 cm in 3 m to a drain and be made non-slip. It is important to have the interior walls strongly constructed to withstand occasional rough usage and be finished with an impervious surface to permit repeated cleaning.

If a line of loose boxes is to be built, consideration could be given to only extending the dividing walls to 153 cm and having a grille of steel bars up to the ceiling. These can be of 2 cm diameter at 7 cm centres. Divisions between boxes may be of timber, but if so a steel plate 30 to 46 cm high by 5 mm thick must be put along each side at floor level to prevent damage from kicking.

The door is placed to the side of the box and must be at least 1.2 m wide and 2.3 m high. If the entrance opens directly to the open air, the doors are always in two leaves, the lower half being about 1.37 m in height. They must open outwards and should then swing close to the wall to cause no obstruction. They must be of robust construction and the lower door must have either a metal bar or sheet metal fitted along and over its upper edge to discourage crib-biting. A stout bolt is fitted to the upper leaf and two bolts or catches put on the lower leaf.

Each loose box must be fitted with a trough to hold cereals or concentrates and a hay rack. Water may be provided either in a small automatically filled water trough, or by bucket. But if the latter is used it is essential that it is firmly held to the wall by metal hoops.

Stalls

In stalls, heavy horses are allowed 4 × 2 m (13 ft × 6 ft in) while for hunters, 3 × 1.8 m (10 × 6 ft) is sufficient. Divisions either wholly of wooden boarding or with an open barred top and wooden bottom are made either between each horse, or more commonly, between pairs of horses, Each stall slopes from its front, having a width of 2 (6 ft 6 in) m to the back, where it is 1.37 m (4 ft 6 in). The post at the back holding the stall division has been known as the 'heel post'. In front of the stalls there is often a passage to permit feeding into the mangers. Each horse can be secured with a

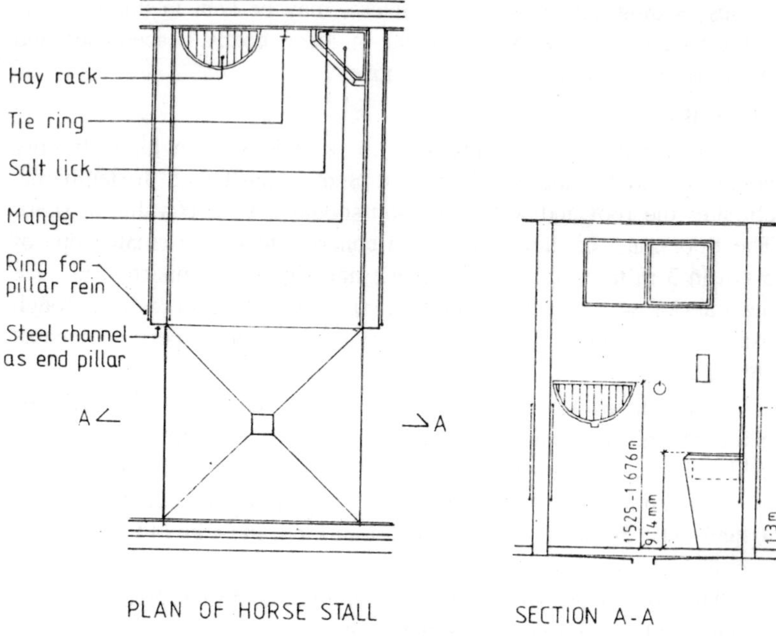

Fig. 2.4. Horse stall.

neck strap from which a rope passes down through a hole in the manger frame, on the end of which is hung a weight sufficient to keep the rope taut, or to a ring bolted onto the wall.

A space of 2.4 m (8 ft) must be allowed behind each stall if there is a single row, or one twice as wide if there is a double row of stalls. There should be a hay-rack incorporated in the manger.

A channel for drainage must be constructed behind each stall. With a single row, the usual dimension is 20 cm wide × 8 cm deep.

Floors

The floor of a stable or loose box must be constructed of durable material, be easily cleaned and not be smooth and slippery even when wet. In times past, many stable floors were laid with Staffordshire blue bricks. Today most are made with cement of at least 60 mm thickness. Its surface must be carefully roughened to prevent slipping. Fairly frequently ridges in a herringbone pattern are found, but is doubtful if the small channels so formed to take the urine to the drain are effective because of blocking by dung.

There must be adequate lighting both from windows and electricity. Windows must be sited above the horse's head.

Drainage

The removal of urine and water used in washing down the floors of stables or loose boxes must be carefully considered if foul air or excessive humidity is to be avoided. The slope of the floor to the drain should be 10 mm per 60 mm and in stalls from front to back.

In stalls there should be a gully in the centre of the 2-m space behind the standing area. In a loose box, it should be in the centre. It has been observed that most horses will urinate ('stale') in that part of their area where bedding tends to accumulate. It is important to have a wire grid over the gully, which should be untrapped (it is undesirable to have urine held in the trap). This grid should be removed and the drain cleared every time the box is cleaned. The outlet should lead by a short branch to the main drain which must be trapped before entry to any larger drain. It is recommendation that every six boxes have a ventilation pipe from the drain to a high enough level. All gullies should be flush to the ground and hinged.

In an attempt to save labour costs, some establishments have tried a deep litter system, thus making the provision of drainage unnecessary. There is then an insuperable difficulty in reverting to the more traditional system of removing urine and bedding should this ever be required.

Ventilation

Natural ventilation has been the common method used, although the availability of reliable variable speed fans that may have heating elements incorporated has, in places experiencing very cold winters, offered the possibility of using a mechanical ventilation system.

Two criteria must be met if the ventilation of the stable is to be satisfactory. There must be no draughts and there must be a sufficient rate of air flow. The latter is measured by the rate of air change each hour, and must therefore relate to the size of horse being kept and the dimensions of the loose box. An average hunter or riding

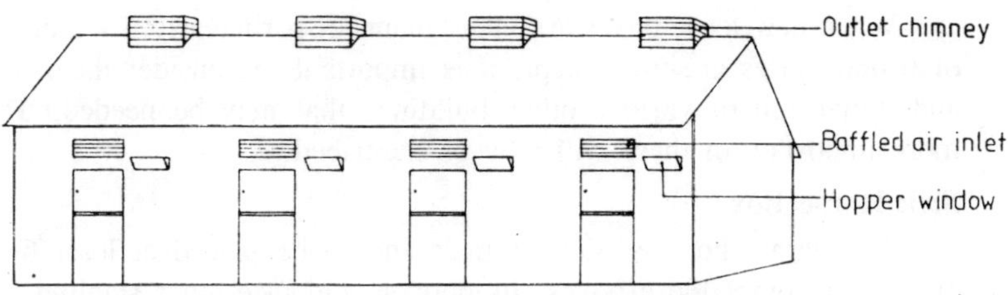

Fig. 2.5. Ventilation for horse loose boxes.

horse of moderate size is likely to weight approximately 450 kg and will require 28.3 m^3 per hour. It is accepted that eight changes of air per hour represents a maximum and that three per hour is a minimum.

The difficulty is to make the right assumption over the wind effect resulting from the situation of the building. If it is in an exposed position, there are many parts of the northern temperate region where the average wind speed is in the region 14-18 km per hour; while in sheltered places it may be as low as 1 km per hour.

Draughts are avoided by arranging for the outlet of air to the stable or loose box to be in the ceiling or, if in the wall, well above the backs of the horses. The simplest method of providing ample ventilation is the use of hopper-type windows with side cheeks that are adjustable. These must be above the level of the horses' heads, and if there can be a similar window on the other wall it should not be placed directly opposite.

The normal practice is to have stable doors or loose box half-doors open during the day and only closed at night. If there is a likelihood of having doors closed for longer periods, the removal of stale air by means of utilizing the 'stack' effect and taking the exhausted air out through roof ventilators should be considered. The regular flow of air results from the well-known fact that heat moves air upwards. If the size of the entry and exit are correctly expressed, there will be a steady flow of fresh air.

The air flow resulting from the stack effect can be calculated and is found to function satisfactorily if the ratio of inlet to outlet is at least 2 : 1. A method that can be used to calculate the sizes and outlets relates the rate of air flow (m^3 per h) needed by each horse (taken to weigh 450 kg) with the mean wind speed (km per h), using a constant varying with the ratio of inlet to outlet. Taking this as 2:1, and the height of the outlets above the inlets as 4 m, and a 10°C difference from outside to inside, there will be a requirement for an outlet area for each horse of 0.05 m^2 and an inlet of 0.1 m^2.

When new loose boxes are being planned, particularly if a number of brood mares are to be kept, it is important to consider the siting and dimension of various other buildings that may be needed. The more important of these will now be described.

Sick Horse Box

A separate box for sick animals should be placed at least 6 m away from other loose boxes. Its door should face other stabling, so that any horse confined due to injury or illness need not feel entirely cut off. Although the placing of a sick or injured horse in a sling is not nowadays a frequent necessity, it is advisable to incorporate a beam capable of withstanding a weight of up to 1.5 tonnes, 3 m above the floor. Heavy horses may weigh more than one tonne. Feed trough, hay rack and water bowl should be placed as near the door as possible. It is desirable to have this box about 50 per cent larger than the normal loose box.

Feed Room and Barn

The feed room is conveniently located next to and even contiguous with the feed store and barn for hay and straw storage. It should be in a central position along the line of loose boxes. It has to contain the bins in which are kept cereals, bran, oil seed meal, and any nuts or pellets. There should be shelving on which measuring equipment may be placed, a suitable set of scales, sink and draining board. The floor should be smooth, have an adequate fall to a gully and be suitable for frequent washing down.

The open barn for the storage of forage, should be large enough to hold a sufficient amount of forage of suitable quality to last the entire winter. There must also be space to store cereals and any other ingredients of concentrates that may be home-mixed in bulk form.

Saddle or Tack Room

It will also be plain that where even a few horses are kept, there is need of a room for keeping saddles, bridles and other harness and equipment (commonly called 'tack'). The precise size must depend upon the number of horses kept in work, but it should have enough space not only to store the tack, but also provide an area where all used tack can be properly cleaned.

This section, with sink and stands for saddles, hooks for holding bridles and reins as they are cleaned and a drying frame for saddle cloths, should be entered directly from the stable yard. Some may feel that there is an advantage in having this as a separate room, from which one proceeds into the tack room proper.

Hot Climate Housing

In semi-tropical and tropical climates where Thoroughbred are kept, or there are establishments for other breeds, there are special problems associated with the heat gain within buildings during the day and its dissipation.

The enclosed quadrangle arrangement favoured in European and North American designs for the housing of horses is not appropriate,

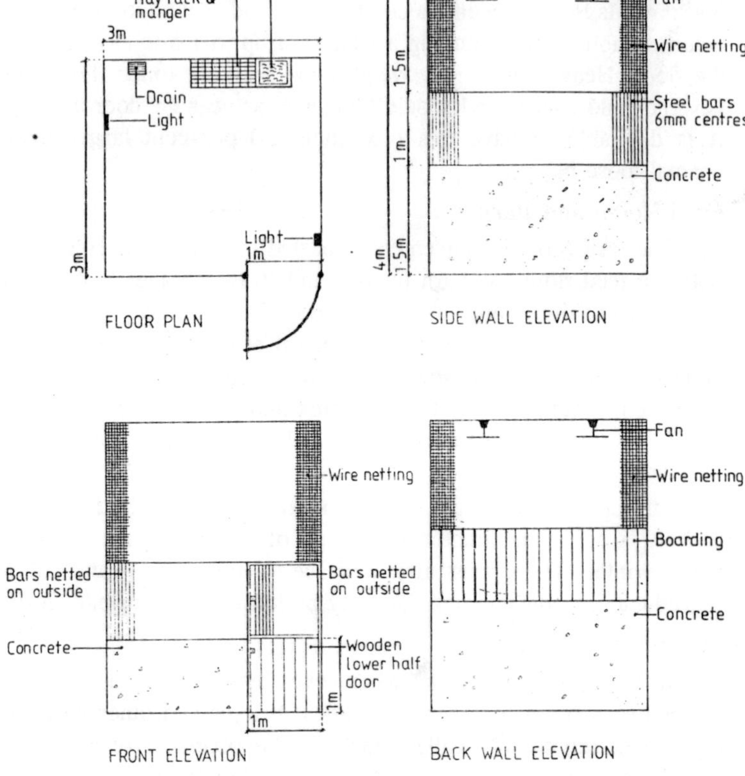

Fig. 2.6. Design for tropical-type horse loose box.

for it is necessary to site the line of loose boxes so that the prevailing wind can be used to the utmost extent.

Two additional points have to be considered: the height of the ceiling should be at least 4 m and, according to whether the location is the northern or southern hemisphere, the side that takes the sun should have a larger overhang of the insulated roof in order to shade as much of the wall as possible.

Ventilation can be provided by only taking the walls to 2 m in height and using bird-proof netting to the eaves. On windless days, or if the humidity is high, the use of ceiling fans may be necessary.

Control of Flies

Wherever stables are built, attention should be given to reducing worry to horses by biting or stable flies. Mechanical means of distribution of insecticides are available and may be used with the synthetic *pyrethrins*, at present very effective. Electrically operated fly traps are also effective. In addition, the regular and frequent cleaning of stables and yards must be undertaken. All faeces should either be taken straight away from the property or placed in a manure pit, where regular spraying against developing fly larvae must be carried out.

Fire Precautions

It must also be remembered that if loose boxes are built of wood, there exists a considerable fire hazard. There must therefore be provided an ample supply of water and suitable equipment at a central point. It also emphasizes the need to have bolts and catches on stable doors that are kept in smooth working order so that, in the event of fire, horses can be speedily released from their boxes.

Cattle

Dairy Cows in Temperate Climates

In Europe, there are still a large number of smaller milk producers who make use of cowsheds or cowhouses in which to overwinter their milking cows and in which to milk them throughout the year. These are nearly always old buildings, and as the relatively rapid change from a large number of dairy farmers with small herds to fewer persons milking many more cows, there has to be a change to a system to covered yards for feeding and either cubicles or kennels for the cows in which to spend their time except for milking twice daily, or occasionally, a thrice daily milking in a milking parlour adjacent to yards and cubicles.

Cowhouse

If a cow has to spend all her time in a stall yoked or tied by a neck strap, feeding and lying in the same place, as well as being milked there, it is clear that floor, walls and the sides of the stalls must be capable of frequent, careful cleaning and the drainage must be effective. Depending on whether there are one or two lines of stalls in the house, so determining its width, will decide whether roof lights, i.e. glass or clear plastic panels, are necessary.

Ample light, both during daytime and at night and early winter mornings is essential. This will be achieved with daylight if the total area of glass is at least one-tenth of the floor area. If there is only one line of cows, the main windows should be behind them in order to illuminate the animals' hindquarters adequately. Electric lighting must be fully adequate.

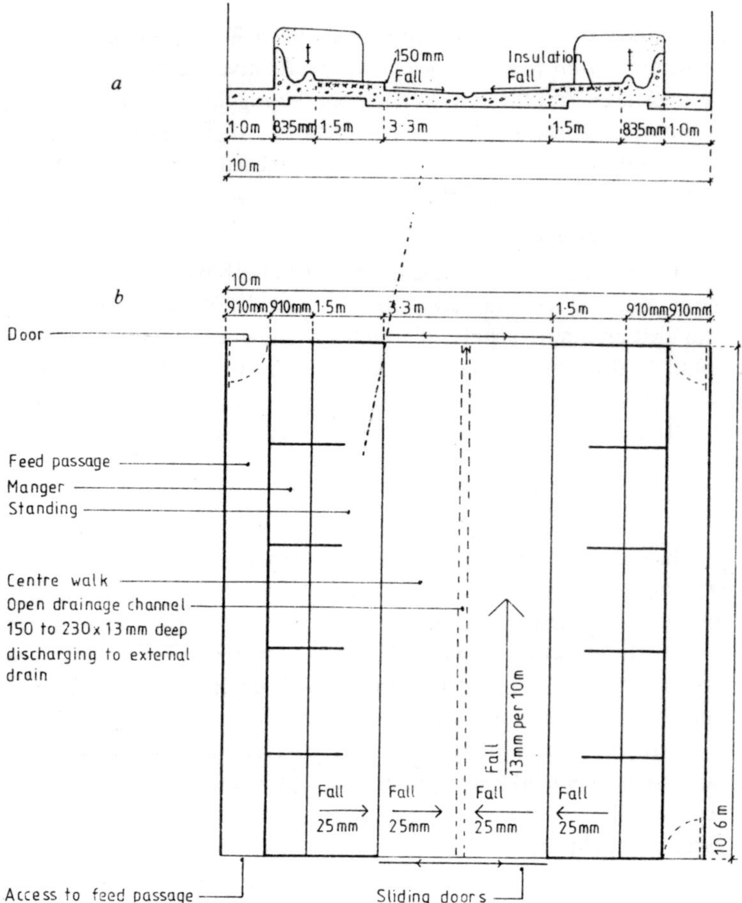

Fig. 2.7. (a) Double-row cowhouse. (b) Floor section with central drainage for cleaning with scraper blade.

The method used for removing solid excreta will determine the width of the space behind the cows: whether a chain or cable-pulled scraper, or a tractor-mounted one, or a trailer taken in for direct loading. Washing down is best done by a pressure hose.

Milking Parlour

This forms an integral part of the commonly used loose-housing system where cows are kept off the fields for part of the year. The total daily movement of the herd must be considered if frustration for men and cows and lost time are to be avoided. The important movement

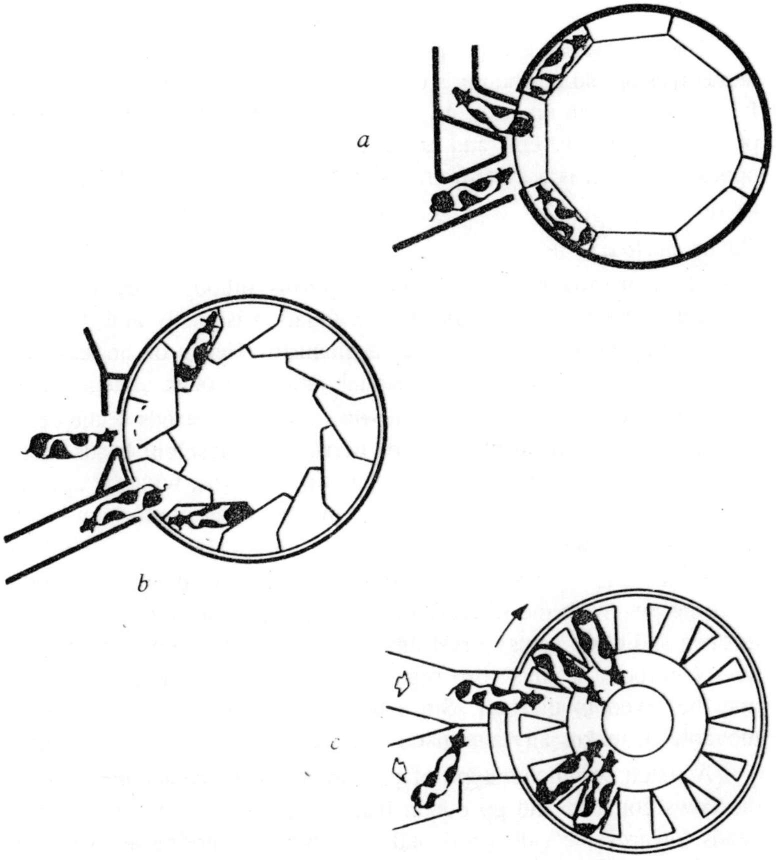

Fig. 2.8. Types of rotary parlours. (a) Tandem—operator inside. (b) Herringbone—operator inside. (c) Abreast—operator outside.

is that of getting the cows to and from the milking parlour speedily and calmly. There must be yards for their collection and passage-ways for entrance and exit. Linked with the latter, there must be drafting arrangement to take any cows that have to be mated or artificially inseminated, or have veterinary attention. For the latter purpose, there will need to be the incorporation of a suitable crush.

There are a number of ways in which cows may be placed once they are got into the milking parlour. Most commonly, the head of each cow is placed towards the outer wall, with the person in charge of the milking operation in a pit behind the cow to save having to bend down.

The commonest exceptions are the abreast and the rotary parlours. Sometimes an old cowhouse is converted to take cows only for milking. It may have them entering on one side, going on to a level or raised platform (about 40 cm) and going out afterwards from doors on the opposite side. It is a system only suitable for small herds looked after by one man.

Rotary parlours

These are found on relatively few farms although they have been available for nearly 50 years. The installation is costly and there has to be constant servicing of the mechanism. Herds of at least 100 milkers are needed to justify the capital outlay and other costs involved.

There are two main types. The operator either stands in the centre with the cows held in a tandem or herringbone fashion, heads facing outwards, or they have their heads pointing inwards, with the operator on the outside. More cows can be got on to the platform with the herringbone arrangement.

With the *tandem* and *shute* type of parlour, the cows stand parallel with the pit. With the former, there is the necessity for the attendant to open and close gates to restrain each cow as she comes in and goes out. With both he must also regulate any feeding of concentrates that may be given in the bail. Since the gates in front of the cows are movable, it makes any automatic arrangement for feeding difficult.

A much commoner type of parlour is the *herringbone*, in which the cows come in and go out in batches, standing to be milked with heads towards the side walls and their hindquarters close to the pit. They stand echelon fashion, thus making maximum use of the parlour space.

Herringbone milking parlour

If concentrates are to be fed in the milking parlour, either remote or manual control by the person in the pit, or electronic operation by the animals themselves will permit the feed to be delivered into a manger at their heads. The hoppers containing the concentrates are normally filled from bins stored in the space between ceiling and roof.

There has been a pronounced swing away from giving any supplementary feed in the milking parlour. The rate of eating varies between cows, with an average for pelleted feed of 0.42 kg per minute, and for meal of 0.29 kg per minute.

The practice of feeding cows during milking is a very old one, associated with the belief that cows would not readily come in from pasture to be milked and would be more content during the milking process if they could anticipate receiving palatable food in the bail. This was when cows were milked by hand with varying degrees of skill and, thus, of speed. The modern milking machine extracts milk at a rapid and constant rate. This fact, along with the massive world-wide shift towards the relatively placid deep-milking Friesian/Holstein breed. and the ever-increasing need with large herds to expedite the milking routine, has brought into question the assumption about feeding during milking. It has been found that when no concentrates are given the amount of milk is not reduced so long as this is the system to which the newly calved heifer is introduced. With older cows, there is a varying time of adjustment.

Milking bail

Originally a system designed to permit milking near the cows' grazing area and where the climate and soil structure were so favourable as to avoid the need to keep the herd inside during the winter. The original Hosier bail was relatively small and was on skids to make it movable. There must of course be an attachable water supply. Electricity is provided by a separate generator if the bail cannot be connected to a mains supply. Where there is to be a permanent siting of the bail, a concrete base with adequate drainage must be laid, with an attached dairy room with refrigerated bulk tank.

Housing and Yards

Where dairy cows have to be brought inside for the 150-200 days of the winter months in the northern hemisphere, the usual accommodation now being provided is collection of cubicles or kennels in which they will spend most of the night.

Kennels

The difference between cubicles and kennels is that the latter are only roofed over the actual part occupied by a cow while she is lying down. In order to obtain the maximum shelter from cold winds and driving rain or snow, it is customary to build kennels in double rows.

Cubicles

It is most convenient to incorporate cubicles within a building that includes the feeding space, and sometimes access to a strawed 'loafing' area and/ or the silage face, if an ad lib system of forage feeding is adopted.

The dimensions of the cubicles are determined by the breed in the following manner:

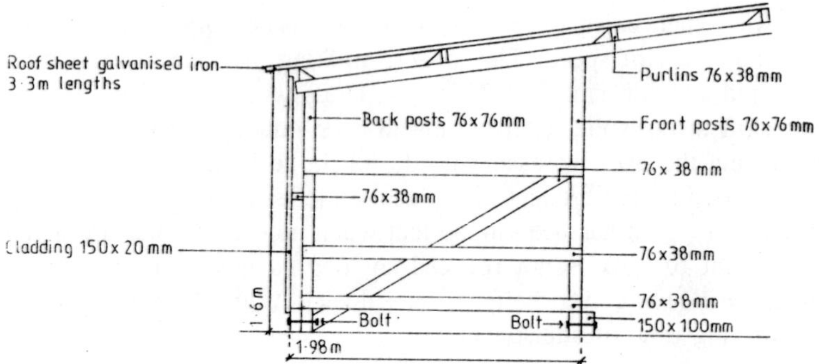

Fig. 2.9. Cow kennel with wooden frame + metal roof.

	Length, m	*Width, m*
Friesian/Holstein weighing over 650 kg	2.08	1.12
Ayrshire weighing 400-650 kg	2.0	1.10
Channel Island weighing less than 400 kg	1.8	1.0

If cubicles are arranged on both sides of a passage, this should be 3.0 m wide.

Floor and bed

The floor of the cubicle or kennel is raised about 25-28 cm above the passage way. A base of hard core or limestone well rammed is first put down. This may be finished with a layer of fine (3 mm) chalk or have concrete laid on top made slightly concave, Peat, sawdust, wood shavings, heather, sand or a rubber mat may be put on top. Straw or shredded newsprint make good beds. There must be a slope of 7.5 cm from front to rear where there is a heelstone or ridge to help keep in the bedding. If the cows are to be comfortable and infection or damage to udders avoided, there must be regular attention given to removing any faeces and, if necessary, removing the bedding.

Sides

There are a great many designs for the divisions between cubicles with the height varying between 100-120 cm. If of tubular steel, a single upright on a level with the cow's rump will be sufficient, polypropylene rope stretched half-way between the upright is satisfactory if tensioned with a 6-mm rod. Any horizontal building rail or batten should be 45-50 cm from the floor.

Feeding

Feeding is most commonly from a line of open troughing on either side of a central passage wide enough for a trailer, feed wagon or other mechanical device to operate to distribute the ration. This may consist of silage or hay as the source of forage, either fed separately and the concentrate placed on top, or fed already mixed.

Modifications of this arrangement are: (1) self-feed silage, in which the cows are permitted to feed ad lib at the silage-face; (2) division of the cows into either three, or at least two groups, high and low producers and dry cows, which are then given access to a complete ration that is balanced for energy, protein and minerals appropriate to their level of production.

Movement of Cows

The effectiveness of any housing unit for dairy cows will be shown by the ease with which the twice or thrice daily routine of movement in and from the milking parlour is completed, as well as the cleanliness and general air of contentment that comes from a herd of well-fed, rested and quietly managed animals.

There should be a space allowance of 5.5 m^2 for each cow of Friesian size wherever the cows are allowed to move freely. If feed is put out in a manger or trough two or three times daily, there should be a space allowance of 50-60 cm in range from a jersey to a Friesian. If a complete ration is available from a feed wagon containing a full day's ration, a smaller feeding space is sufficient since cows can feed at any time.

Collecting Yards, Race and Crush

The shape of collecting yards alongside the dairy varies from rectangular to circular but an average adequate space allowance is 1.4 m^2/cow. The entrance gate must be the full width of the yard and, in long rectangular yards, is placed opposite the dairy entrance.

A circular yard with a mechanically controlled backing gate helps to keep the cows moving to the dairy entrance. A gate of radius 8.5 m is able to contain up to 150 cows in three quarters of a circle. From the exit of the milking parlour, there should be a gate that can be opened to permit any cow needing veterinary attention to be separated and taken to pens for insemination or to a crush.

Ventilation in dairy cow buildings

Natural ventilation is always used, the aim being to provide ample draught-free air movement by having space boarding from the top of a base of building blocks, brick or concrete, having a height of 2.30 m. The space boarding to the eaves can have slats of 125 mm width and a gap of 25 mm. An opening along the roof ridge is desirable.

Dairy Cows in a Tropical Climate

Potentially high-producing cows have to be provided with shaded areas in yards, in addition to the usual facilities of getting them to and from the milking parlour.

Because of the unavailability of pasturage, notably in the dry tropics and in many subtropical situations, it becomes necessary to keep all cattle in open yards throughout their lives. Such yards may vary in shape and size according to the number of cows in a particular age or production group, but a minimum area of 25 m^2 per cow should be allowed. Growing stock need only 15 m^2.

It is necessary to provide shade over areas where the cows feed and drink and over a resting area. In the humid tropics, attention must be given to the surface of the open yard, its drainage and the removal of slurry.

In the dry tropics, the many months of high temperatures emphasize the importance of selecting the most appropriate roofing material and of having the roofs high enough to reduce radiant heat yet provide enough shade. There is not complete agreement amongst those with experience of dairy farming under these conditions whether it is best to site the shade along an E-W or N-S axis. With the prevailing wind often coming from a northerly direction, an E-W axis allows full use to be made of its cooling properties. Although units have been put up with side walls, experience in many parts has shown that entirely open yards are better.

An exception is where the funnelling effect of any prevailing breeze can be utilized by having sides part-way to the roof over some 'loafing' areas and leaving the ends open, one of which points to the prevailing wind. One unit inspected by the author that had such a building in one of the large yards was much used by the cows because it was cooler.

Schemes have been tried whereby sides have been put on the feeding area and fans used to draw in air over wet sacking hanging from the roof over gaps in the walls to the outside. It may be combined with turning on a fine water spray directed on to the backs of the cattle as they feed at the trough. There is no indication from production figures that such measures have a beneficial effect to justify the increased capital and maintenance costs.

What seems clear is that potentially high-producing dairy cows of most breeds brought into the tropics from temperate climates go through a period of acclimatization lasting at least one year. The majority adjust to the high temperatures of the hot season and attain a production level of between 60 and 75 per cent of what they would probably have reached in the environment from which they came.

The commonest practice is to import in-calf heifers into tropical countries where dairying is developing, thus making the test of production during the first lactation a severe one. If ample shade is provided and such heat-holding material as concrete avoided in roof construction, experience indicates that entirely open yards with free access to shaded water and feeding troughs is both the cheapest to construct and the most effective in operation.

The milking parlour must be well ventilated, as well as having a high roof made of material with a low *U* value. The aim should be a thermal value of 1.14 W per m^2 per deg. In addition to having the sides open from above 2 m, but enclosed with fine netting to keep out flies, it is advisable to have ceiling fans at approximately 3 m centres. Any of the standard arrangements over collecting yards and facilities for handling cows may be used.

Bull Pen

If bulls are kept, their pens must be sited in a suitable position in relation to the normal movement of the cows to and from the milking parlour. Time and trouble in transferring cows in season for mating must be kept to a minimum. In a tropical country, the shaded area for the feeding and water troughs should be adjacent to an outside fence to avoid the need to enter the yard. This shaded part can be extended over a fenced service area where stocks and a crush can be sited, so that any needed attention can be given.

Calves

Unless driving rain during the wet season or cold winds make an entirely enclosed shed necessary, an open shaded area is likely to prove most effective. It may be protected from cool weather winds by a temporary wall of straw bales.

With exotic breeds of dairy cows in tropical countries, it is customary to follow the usual practice of removing calves within 1 or 2 days of birth and rearing them on whole milk followed by milk-replacer until they can be weaned at 6-8 weeks.

If separate pens are preferred, an effective arrangements is to use weld-mesh for the sides of each pen and to use straw bales to separate the pens. Feeding and watering is by bucket held in brackets at the front. Each pen should be 1 × 1.5 m and the whole area shaded by a roof with high insulation.

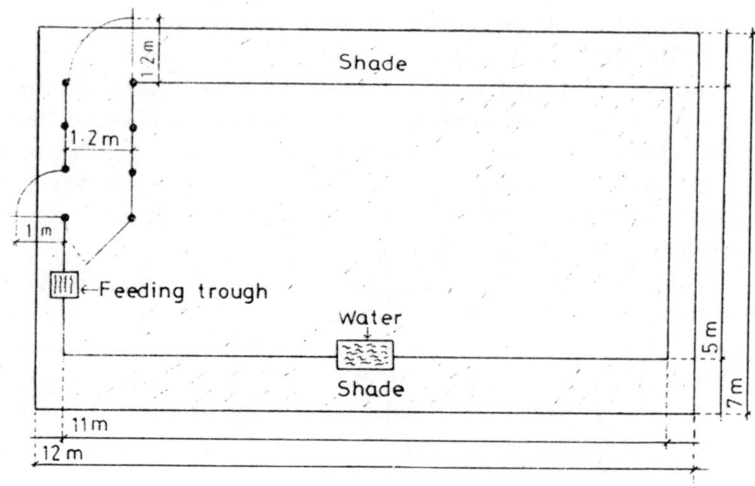

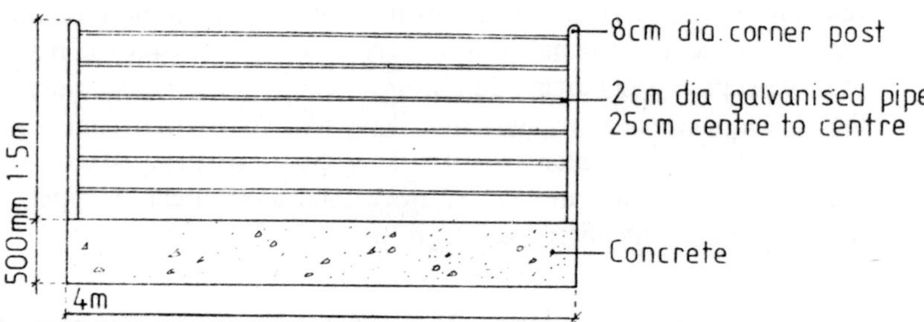

Fig. 2.10. Bull pen for the tropics; detail of yard railings.

Heifer replacements can be effectively reared in shaded yards similar to those for cows except that the space allowance can be reduced to 3.5 m^2. Trough space should be 5.5 cm for Friesian-type heifers.

Beef Cattle

In the U.K. and in other European countries, most beef is produced under three systems: single suckler, semi-intensive and intensive.

Single suckler system

Beef-type cows are often kept in groups planned to calve either in the autumn or spring. If it is necessary to house cows and calves, the most satisfactory way is to provide sufficient space in a well ventilated house with plenty of straw for bedding. If they are calving in the autumn and must therefore be housed for approximately 180 days, up to 1 tonne of straw per cow will be needed. They will need between 60-84 m^2 per cows. With a spring-calving herd, the space allowance is reduced since the calves will be small until able to go out to pasture in the spring. If the beef fattening unit is in an area where cereals are not grown and the cost of straw is too high, it is possible to keep the cows and calves on slatted floors with an area of solid flooring near the feed troughs. A slatted floor is not satisfactory for cows in milk because of the transfer of faecal contamination on the teats and the danger of mechanical injury to the udder.

Semi-intensive

This method aims at having cattle at marketable weight by 18-24 months and involves at least one winter during which the animals will be at the growing/fattening stage. If they are autumn-born calves, they will be yearlings entering their second winter; if spring born, they will be spending their first winter indoors. The space allowance will depend upon their weight. It is generally agreed that for the 450-540 kg beast of 18 months, a bedded area of 4.0-4.6 m^2 per animal is desirable, but on slats this can be reduced to 1.8-2.0 m^2. The feeding trough space should be 600 mm.

Intensive method

This aims at steady weight increases of close of close to 1 kg/day and depends upon a very high proportion of cereal in the diet. In the U.K. barley has been most commonly used. The cattle stay inside for the whole of the 12-14 months it takes them to reach killing weight. As a result of the much increased prices of cereals in the 1970s, along with higher prices of suitable calves, the margin of profitability for this type of beef production was so reduced that it has relatively few adherents.

Beef Cattle Sheds

There are three types of housing used on beef producing units in much of Europe: (1) *calf-rearing*; (2) '*follow-on*' or *grower pens*; (3) *fattening* or *finishing* pens. The majority of home-grown beef comes from the unwanted calves produced by dairy farmers. The management and housing of calves up to weaning at 5-7 weeks, destined for beef, is similar to those who will become dairy replacement heifers.

Grower pens

Whether calves are reared in an environment-controlled house or not, they are likely to have been kept inside at a temperature above that to be found outside during the winter months. The requirement of a house where the young, weaned animal will spend the 6-8 weeks following weaning needs to be well bedded with straw, or other suitable material, and must aim to provide conditions that permit the young animals to 'harden-off' in atmospheric temperatures in as comfortable conditions as possible. In northern latitudes this means facing the main opening, often an entirely open side of the shed, towards the SE. If existing buildings are used, care must be taken to see that ventilation is really adequate. The space requirement is 1.8-2.7 m^2 if bedding is used, with 45 mm per beast of trough space.

Fattening house

Monopitch open-front. Such a building should face to the south and never be built in an exposed position. On the other hand, it must not be hemmed in closely by other buildings or the ventilation in windless weather will be inadequate. The depth is up to 8 m with feeding trough close to the open side. If the building has a solid floor and litter is used, it could be used for 'hardening' the calves and a proportion kept on for fattening. The space allowance must be increased to 4.6 m^2 and the trough space 600-680 mm.

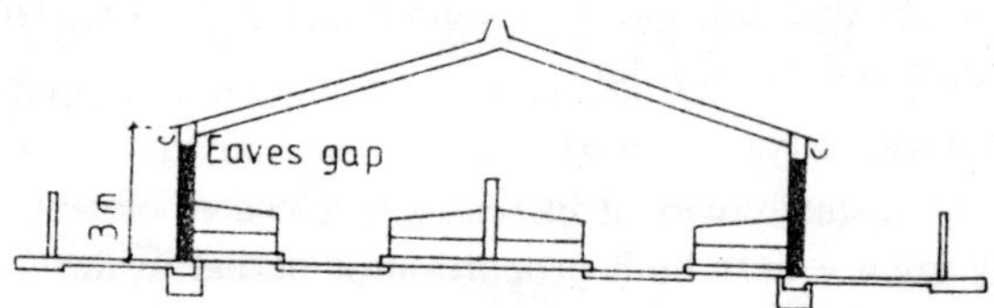

Fig. 2.11. Typical beef fattening house.

Portal frame. This is the commonest building used nowadays. It has a central passage of at least 2.7 m to allow tractor and trailer to bring in feed. If a complete ration is mixed in a forage box, the passage may have to be wider. Some have found that the use of central conveyor-belt system, although expensive to install, can be justified as a result of the saving in building costs that result from eliminating the central passage. There must, however, be a walk-way to allow the cattle to be kept under observation. The central feeding passage is sometimes raised to allow for the winter build-up of dung where there are solid floors. This helps to make observation of the cattle easier.

Very large clear-span buildings are possible, but a minimum span is 9 m. The wall height should be 3 m to the eaves, but it is customary to have space boarding or an eaves gap of 450 mm. There should be a ridge gap to let stale air escape, to which an apron flashing should be attached to prevent rain being blown up the roof and into the building.

Slats

The cost of straw and other bedding in parts of many countries has forced beef producers to make use of slatted flooring. This also means that careful provision must be made to deal with the slurry that is produced.

Concrete is now almost always used for the slats, which may vary slightly in shape. The width of a suitable slate for cattle must be 100-125 mm and its top should be slightly cambered and the edges rounded. It must taper to help the passing of excreta and there must be a gap of 3.8 mm between slats.

The importance of adequate ventilation in slatted houses must be emphasized. If slats remain wet, they are slippery and there will be swollen hocks and knees as a result of falls. There should be limited storage of slurry under cattle standings because of the danger of crust formation and the liberation of toxic gases (methane, carbon dioxide) when being cleaned out.

It is better to have no more than 1 m space under the slats, with a sharp fall (1 in 40) to a slurry tank as close to the end of the house as possible. How and where this is emptied will depend upon the size of the unit, the nature of the soil and size of the surrounding holding of land. It may be taken as a rough guide that fattening cattle will produce 0.028 m^3 slurry per animal per day. On sloping sites where stock access can be at the higher level and slurry may be difficult to dispose of, the slurry can be collected in a cellar excavated on the low side of the ledge; the liquid is allowed to drain away and the solid material removed by tractor with bucket or fore-end loader.

Money can be saved on building for beef animals by providing topless kennels and feeding troughs outside with only the trough itself being covered. The conditions are not comfortable for the stockman (an increasingly important consideration in deciding upon the type of housing to be put up) and there is a large quantity of diluted slurry (from rainwater) to deal with.

Bull Beef

Slowly the meat trade's prejudice against handling bull beef diminishes and the future may well see more fattening of entire males, so to make use of their greater food conversion capacity in the production of leaner carcasses.

The Ministry of Agriculture's code of practice for buff beef premises recommend that a stout fence surround the area, so that there can be no chance of an animal getting free, and that all animals be fed and watered without the need of the attendant to enter their pens. Any pen that needs to be entered should have an emergency escape route. Gates, posts and rails must be stoutly constructed. Animals of the same age and size should be kept together

It is claimed that slatted floors are very suitable for bull beef fattening because the slight uncertainty of footing tends to keep the young bulls more docile.

SHEEP

Many sheep farmers, both in the lowlands and uplands of the U.K., who have changed to housing their ewe flocks during part or all of the winter, or at least during the spring lambing, are convinced that reduced losses of both ewes and lambs and, in heavy soil areas, the saving of pasture from treading, justify the expense of providing the housing and any extra labour involved. It is important to keep building costs to a minimum. They may be avoided altogether by using hay or silage storage areas in a cattle building at the end of the winter if the flock is only brought in for lambing. Temporary pens are constructed, often by using hurdles or weldmesh and straw bales.

Lean-to buildings

Sneep are prone to respiratory infection if kept in enclosures without very ample ventilation. For this reason, as well as a means of minimizing costs, open-sided shed are frequently built. They should have a southerly aspect and an overhang over a concrete path to make feeding and observation of the sheep a pleasanter task in bad weather.

Portal frame building

If a frame building is to be erected, it should have an eaves height of 2.4-3.0 m and a minimum span of 9 m. The walls may be of light construction, single brick blocks, metal sheeting, oil tempered hardboard to 900-1000 mm, above this being open cladding. The roof may be of asbestos, aluminium or metal sheeting with a translucent roof sheet for every bay to provide daylight. If open boarding is not provided, the roof ridge must be open.

The flooring, unless straw is plentiful, should be of soft wood slats. These can be laid upon splayed joists 100 m × 50 mm on a concrete bed with piers and in sections of 2.4 × 1.2 mm to permit once-yearly cleaning. The slats themselves should be 32-37 mm wide

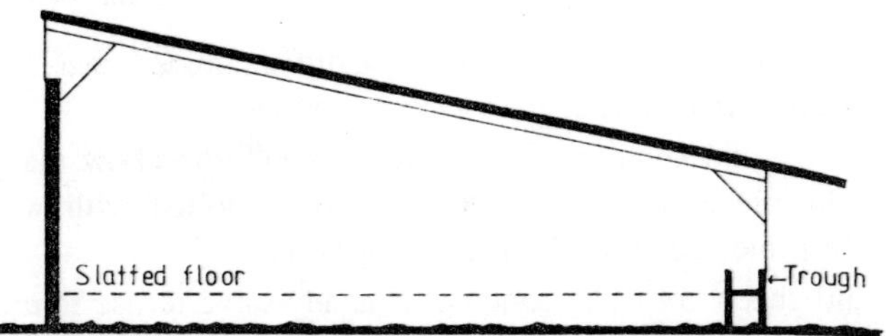

Fig. 2.12. Open-fronted shed, slatted, with overhang, to keep feed and attendant dry.

× 36 mm deep and with a 15 m gap. They should be laid parallel to the door opening, giving the sheep the illusion of a solid floor as they enter. The doorway should be 2.4 m wide to permit entrance of a tractor for cleaning.

The space allowance varies with the size (and breed) of sheep in the flock. For large ewes with a lamb 1.3 m^2 is necessary; for smaller breeds 1 m^2 is enough, assuming slats are used. If straw bedding is used, the space allowance should be increased by a fifth.

Its is customary not to keep more than 10 ewes in one pen. The smaller the number kept together, the more certain can the stockman be that all are getting sufficient food. Hay is usually fed in racks (150-200 mm per ewe) on either side of pen partitions and concentrates in troughs (400-450 mm per ewe). A combined rack and trough known as the 'Scandinavian feed box' has recently become much used. The box can fit between the bars of a pen division and is 470 mm wide × 300 mm deep × 2400 mm long. Concentrate mixture or pellets are put into the bottom of the box, a length of standard weldmesh (150 × 75 m) is placed on top and the hay distributed above it.

Electric light

Buildings in which lambing takes place should have adequate lighting for inspection at night.

Lambing pens

On some properties in hilly regions it is the custom to put up temporary pens of about 0.4 m^2 by using straw bales or hurdles with a wall or thatched hurdles behind to break the wind. In Scotland, circular roofless structures with only one south-facing entrance, made of stone, steel or treated wood called 'stells' have for long offered protection in bad weather for hill flocks.

Sheep Handling and Dipping Yards

It is most important that every property keeping a flock of sheep have adequate yards in which sheep may be drafted, drenched against internal parasites, vaccinated, individually examined and dipped as preventive or treatment against external parasites.

The size and number of pens must depend upon how many sheep are on the place, but every site must be supplied with water and should be protected from the prevailing winds.

If hill sheep are to be handled, it is advisable to use four wooden rails, 75 × 32 mm, with a height of 1.05 m, and the rails spaced from the ground upwards at 150, 175, 200 and 225 mm. Three rails are generally thought to be sufficient for lowland sheep at 175, 200 and 225 mm with a total height of 825 mm.

The basic requirements are :

1. A *collecting yard* into which the sheep are first driven.
2. A *forcing pen* with tapered side with large swing gate to move the sheep either through an open gate into the dip or drafting race.
3. *Dip*. This is rectangular and at the distal end a corrugated ramp at 30° to allow the sheep to walk out of the dip into the draining pen.
4. *Draining pens*. There are commonly two of them, side by side, in order to allow enough draining time. The floors of these pens slope (1 in 50) to a drain to allow dip fluid to return to the dip. A two-way drain will prevent rainwater entering and diluting the dip. For a flock of 350 sheep, capacity of 775 litres is sufficient; for one of 500, it must be increased to 1150 litres, and for flocks of 1000 and over, a dip of 6 m long and containing 2550 litres is necessary.

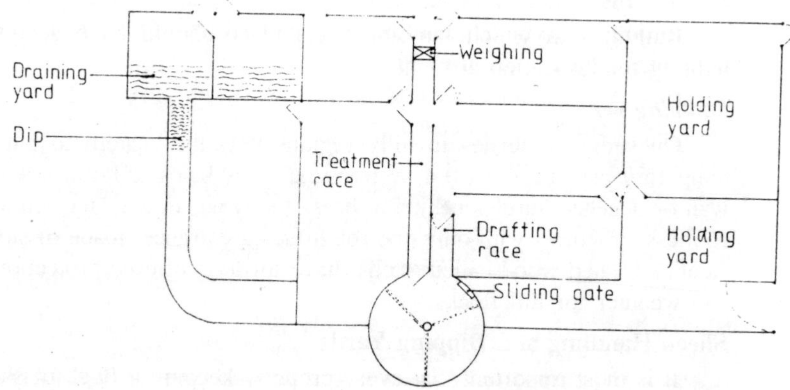

Fig. 2.13. Sheep handling and dipping yards.

5. The other exit from the forcing pen allows the sheep to enter a *drafting race*. This only permits one sheep at a time to go forward. It should have solid boarded sides and be tapered 200 mm width at bottom and 350-450 mm at the top. A gap of 100 mm at the bottom allows the sheep's feet to be inspected and makes bending over the sheep for drenching, etc. easier. The race should be 4-6 m long and at its distal end should have a sliding board or other arrangement to stop sheep going forward, and immediately beyond it there will be a single or double drafting gate to permit sheep to be shed to right or left into separate pens, or allowed to go forward into another pen.

If footrot is a problem, a *foot bath* can conveniently be constructed along an outside fence. It should be 3 m long, with a width of 225 mm and depth of 150 mm.

Sheep have to be shorn every summer and it is convenient, if sheep production is a main preoccupation on the property, to have a proper *shearing shed*. On the majority of European farms, sheep numbers are relatively small and shearing is undertaken in a shed that will be used for other purposes for the rest of the year. Power points are necessary for running the clippers.

It is desirable to have the shearing shed close to the sheep yards so that the ewes may be run off from their lambs and moved in under cover, for it is necessary in a shearing shed that there be cover for 24 hours to have dry fleeces for a day's shearing.

Pigs

The modern improved breeds of pigs possess much the same capacity to thrive on an exceptionally wide range of foodstuffs and adapt to various methods of husbandry as much less selected pigs have done through the ages. It is nevertheless important to remember that the very young pig is unable to control its body temperature and has virtually no hair covering when it is born. In large-scale intensive system therefore heat must be provided for piglets during the first few weeks unless they are born in a tropical country. Because of their large size and slow reaction-time, sows are likely to crush some of the piglets in a litter unless special precautions are taken during and for some days after parturition. This, then, is a critical time for the pig-keeper and there have been a multitude of devices and methods of management tried in order to minimize losses.

It is convenient to consider the housing of pigs in the divisions into which the production of baconers naturally falls, i.e. care of the sow through to farrowing and the weaning of the young pigs, the rearing period and the final finishing or 'fattening' as it was traditionally called.

Sow

Veterinary writers have repeatedly emphasized the desirability from the point of view of preventing disease that there should be no more than 20 farrowing pens in any one house or walled-off section of a house, so that periodic emptying, disinfection and resting of the place can be undertaken.

Dry Sow Housing

With bedding

Where there is a supply of straw, a suitable and simple method is to have the sows in a covered straw yard. Sows will fight and damage each other at feeding time unless they are given an individual feeding space. Thus there must be a line of individual feeders along an internal passage-way or along the outside wall where there is a path protected by a roof overhang. This should be south facing. Each feeder is 500 mm wide × 1.75 m long and has 450 mm trough space. The line of feeders must be built in front of a raised platform to permit the gradual build-up of litter. Each sow can be allowed 3.5 m^2. If a 9-m span building is used, it is advisable to divide it into pens to keep the number of sows in each group to 8-10, a further way to minimize fighting.

Slatted house

Economic pressures have resulted in the need to resort to keeping sows on unbedded concrete floors with part of the area slatted through which dung and urine fall.

For many years the cost advantage of keeping dry sows during their pregnancies in individual stalls has been widely known. Many more sows can be housed within a given area and fighting is avoided. Many patterns of sow-stalls are commercially available, but the essential measurements are that the stall will be 195 m from the rear gate to the front edge of feeding trough, which makes up an extra 375 mm. It will need to be 2.02 m if a glazed tile sunken trough is used along which liquid food can pass.

There appears to be an increase in the use of girth tethers for restraining the sows, thus making a rear door unnecessary. Neck ties are less satisfactory because many gilts object to the restraint and minor wounds result.

When the house contains two lines of sows, it is important that they face each other; an echelon or herringbone arrangement of the stalls is sometimes employed since this permits single pipelines for water and feed to service the two facing lines of stalls. The floors of sow-stalls must be insulated and the top cement carefully roughened to prevent slipping but not to the extent of eroding the sole of the hoof. Rubber mats have been found useful. The front of the stall should be level but a slope in the near half of 25 to 50 mm. The slats at the back are generally concrete; 60-70 mm top width with 21-25 mm gap. Expanded metal flooring lasts longer than slats, but may cause lameness.

There is increasing public concern over the welfare of farm livestock, especially over methods employed in intensive husbandry systems. To keep a sow so closely confined that she is only able to get up, stand and lie down again for practically the whole of her life makes many wonder whether this regime can continue to be justified. A small pen enclosed by railings at the back of the stall to allow limited exercise, although increasing housing costs, is desirable.

A move away from the sow-stall is to a sow cubicle into which each animal can go both for feeding and lying. These may be arranged in groups of 3 or 4 in a double row in the usual 9-m span building with sides to 1.35 m and slatted boarding having 18 mm gaps above to the eaves at 2.4 m. The sows can move about the 2 m passage behind the cubicles where dunging occurs. This should be mechanically scraped. The cubicles are 2.1 m long and 600 mm wide and should be straw-bedded.

Farrowing Accommodation

The sow or gilt due to farrow is moved to a special farrowing house shortly before parturition. Two needs must then be satisfied. The newly born piglets must be protected from being crushed or killed by the sow and they must be provided with a warm sleeping area close by, where temperatures between 21-27°C can be maintained.

Outside farrowing

Very few pig breeders keep their sows outdoors in movable huts or arks these days, because it is a high-cost system in terms of manpower for feeding and moving the huts. It also occupies land that can give a better return in crops. It is said to be a healthy system, reducing the likelihood of cross-infection and the initial cost of buying or building this accommodation is low.

There are various designs, some of rectangular or semi-circular shape, others triangular. All need a baffleboard set at an angle close to the sow's point of entry to prevent wind blowing in directly. There must be a farrowing rail 250 mm from the floor and 250 mm from the wall, and if the litter is to be reared in the hut, a creep area in which the piglets can go for supplementary feeding but which is inaccessible to the sow. This area can be fitted on to the back of the hut and so be easily accessible to the attendant. It is important that there be very good insulation throughout, roof, walls and floor; and if, as in some models, there is a door at the back to permit a person to go in and attend the sow, this doorway must seal completely when shut.

Inside farrowing

Most sows spend their lives inside buildings that have been specially designed for the purpose. This purpose is two-fold. First, the sow must be mated and, on conception, can be kept with other dry sows during pregnancy. As this nears its end, she is moved to the farrowing house and put into some form of farrowing crate where she will give birth to the litter.

Various designs of farrowing crates are in use, all having the purpose of closely confining the sow so that she may only lie in one direction, with the bottom protecting her at sufficient height (250 mm) from the floor to allow free escape by the young pigs. Since sows vary in size and length, this bar is sometimes adjustable, as is the rear door, holding the sow from backing or stepping on a piglet. There seems to be greater use being made of girth tethers for the sow, thus obviating the need for the rear door. The creep area for the piglets immediately to the side of the farrowing crate is approximately 430 mm wide and runs back the 2.7 m length of the unit. Most pigment prefer to cover the creep area with hardboard and to have an infra-red lamp inside to provide the relatively high temperature demanded by newly born piglets of 21-27°C. The floor of farrowing units is insulated except over any slatted area. Slats have been used increasingly, with the crate and creep area covered in galvanized perforated steel sheets until the young pigs are a few weeks old and not likely to catch their feet on the edge of the slats. These are commonly 75-100 mm wide with a gap of 12-20 mm, which may be enlarged to 25 mm behind the sow. Drainage of solid-floored pens must be 1 in 20 to take urine or any spilled liquid feed to drains. A low wall at the rest or creep side of 550 mm high makes observing and gaining access to the creep area easier.

Solari house

This has proved an efficient type of housing for a small unit. It also has the advantage of isolation and thus the ability of controlling disease spread. Each separate pen is 4.95 m long and 1.5 m wide, with the height varying from 910 mm at the back to 2.4 m in the front. The back has no opening, while in the front is a low wall of 1.2 m, broken by the door at the same height. The creep area is behind the farrowing rail and is insulated and heated. The rails are 150 mm from the floor and 250 mm from the walls and may be removed after 10 days, leaving the creep area.

In the latter is to remain for the growing and fattening period, this creep may also be removed.

In areas of mild climate these houses are economical and efficient, but during extreme winter weather the attendant is exposed during feeding and water bowls may freeze.

Weaning and Rearing

The need to have the sow produce as many litters as possible in her productive life, which is likely to be of 4 years, has made early weaning systems common. From the original system of weaning at 8 weeks, there has been a steady move to devising methods of safely and economically achieving 3-week weaning and, with special equipment and care, weaning at 10-16 days old appears to be increasing.

Litters of piglets are not only of varying numbers, but of different size and vitality. Growth and even survival depend upon the piglet having a plentiful supply of sow's milk. Since sows will easily permit cross-suckling, it is advantageous to have litters of approximately the same age kept in contiguous pens to enable the attendant to foster weak piglets on to a sow with a plentiful milk supply.

As the 3-week stage is reached, many experienced pig producers believe it is advantageous to mix three to five sows and their litters so as to form a common 'weaner pool'. Within a week or two the sows are removed and the group of weaners kept in their pen until moved to a fattening shed.

There has been increasing use made of 'flat deck' accommodation for weaners. These are pens of varying sizes to contain 10 weaners (2 m × 1 m) to larger, even two-tiered arrangements, to house 50 weaners. The floor of the flat decks must permit the complete passing of excreta and yet no cause difficulty of movement or damage to the hooves. Expanded metal or punched steel sheets are at present used but an entirely satisfactory flooring is still sought.

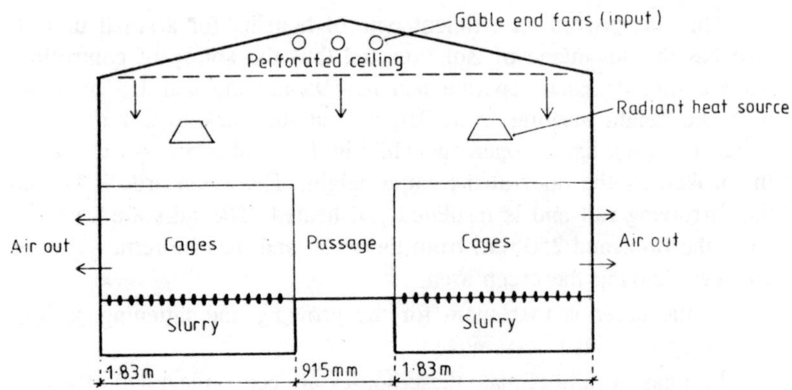

Fig. 2.14. Flat-deck system for weaners; double row, pressurized house.

The temperature and ventilation of enclosed rearing houses must be closely controlled Insulation of roof and walls should be 1.7 W/m^2/deg and ventilation by thermostatically controlled fans linked to adjustable air inlets. Adequate positive pressure will reduce transmittance of infection and prevent smell from the slurry channels.

Those producers who wish to wean at 10-14 days need to use specially heated and ventilated tiers of cages generally enclosed in plastic tents with filtered air.

Fattening House

In many countries the Danish-type fattening house has been used. Pigs of similar size are placed in pens of 3 × 1.8 m to hold 10 of them from 50 kg to bacon weight at 90 kg. Feeding, either in troughs or on the floor, is from a central passage and a dunging area over slats is at the back of the pen or in a passage behind them.

The aim is to have a temperature at pig level of not less than 16°C. In small houses this can be achieved by ridge inlet and exhaust fans at the side exits. It has been shown by the National Institute of Agricultural Engineering in the U.K. that control of the side wall extraction fans should be on an 'on/off' system, linked to automatic adjustment of fresh-air inlets in the ridge. The ridge gaps are connected to a reduction motor and close as the fans shut down.

Modifications made to permit larger numbers of pigs to be fattened in enclosed buildings in order to reduce costs include having one central dunging passage, with cross or side-wall feeding passages. If mechanical dry food dispensers or liquid feeding is undertaken, further space can be saved, with a cat-walk used for inspection.

The generally accepted space allowances for each fattening pig are 0.55 m^2 lying area, dunging area 0.2 m^2 on slatted floor and a trough length of 300 mm when its width is 300-400 mm. There should be a higher level of lighting in the dunging area than where the pigs lie. A 1 in 15 floor surface fall is necessary where there are no slats.

Boar

Many countries have efficient arrangements for supplying semen from progeny-tested boars for artificial insemination of sows by the pig owner. Because the correct detection of oestrus in the sow is time-consuming, large units continue to use their own boars. For the sow to exhibit the outward signs of oestrus and for ovulation to occur, it is advantageous to have the boar within sight, sound and smell (for the pheromones produced by his salivary glands plays a large part in the sow's conditioning). Thus the boar's accommodation should be as close as possible to the sows' and is frequently within their house. He needs a pen 2.4 m by 1.8 m and, if possible, should be allowed exercise outside; a run of 7-9 m is suggested.

Sows on heat are moved to the boar's quarters for mating or are mated in service pens. It is good practice to keep gilts entirely away from boars until they reach 160-165 days of age. A conditioning encounter with the boar then is likely to bring on a fertile oestrus within 7-10 days.

Poultry

There has been a remarkably rapid growth in the use of intensive high-density indoor systems of keeping laying hens and of producing broilers. Less than 4 per cent of the eggs in the U.K. are produced by allowing layers to have an outside run, and a declining proportion of eggs are produced on the deep-litter system.

Rearing Layers and Breeders

Deep litter house

The purpose is to rear the day-old chicks right through to the commencement of laying at about 18 weeks in the same house. Although this must mean that there is waste space for the earlier part of the total period, there is the considerable advantage that the stress of movement to fresh surroundings is avoided, vaccination programmes against Newcastle disease and other infections can proceed to maximum advantage and immunity to coccidiosis gradually acquired.

There are special temperature requirements for the rapidly growing chicken to be met in the first month, normally attained by placing them in partitioned groups under suspended brooders which may be heated by electricity or gas.

Age in weeks	*Brooder temperature, °C*
Day old-1	32-35
1-2	29-32
2-3	27-29

The room temperature should be kept at 15-18 °C for the first 6 weeks. For the first week the chicks are fed upon paper spread under the brooders, but are then given troughs of small size and allowed 2.4 m per 100 chicks. The space and size of troughs increase so that from 13 weeks onwards they are allowed 6.1 m per 100 chicks, and there must be one water point for 100 birds evenly spread over the house.

The space requirements are: 0.025 m^2, 0-4 weeks; 0.09 m^2, 4-8 weeks; 0.18-0.28 m^2, 9-16 weeks. If broiler breeders are being reared, since they are much larger than commercial laying birds, they should be given the maximum space.

The rearing house has a concrete floor, although rammed chalk or earth can be suitable, on which wood shavings, wheat cavings, peatmoss, chaffed straw, or even sawdust or sand, are all spread to a depth of 5 cm and added to monthly. The walls do not have to withstand pressure from the stock housed, so they can be of plywood sheets, matched boarding or similar material. It should have a thermal insulation of 1.07 W per m^2°C, and they are frequently 1.5-1.8 m high. So that the lighting pattern can be controlled—particularly important if the house were to be used for layers or broilers–no windows are put in. The roof may be of board or compressed straw slabs, covered with tarred felt or corrugated asbestos and should have a thermal insulation of 1.13 W per m^2°C. The ventilation, as with all poultry housing, is an important factor in the success of the unit. Most commonly it is by multispeed fans in roof outlets with air drawn in through baffled wall inlets. Nest boxes to provide 12 m per 100 birds are provided. In large units the eggs roll on to a conveyor belt and are taken to the egg room. If eggs are to be collected by hand, the nest boxes must be arranged next to a central passage.

Battery rearing of layers

Although the cost of rearing layers in battery cages is more costly than on deep litter, the advantage of no setback at having to adapt to a new environment at point of lay and the decreased disease risk is making this method increasingly used.

Battery cages are most usually in three tiers in height, but there may be four tiers. The day-old chicks are either all put in the top tier or in both top and bottom if there are four tiers. Paper or fine mesh netting is put over the wire floor of the cage for the first two weeks and shallow trays for water until they can reach and use the nipple drinkers. As the young pullets grow, they are moved into the vacant cages in the other tiers.

The fact of their being in windowless buildings allows the birds to be given a controlled lighting regime. In order to attempt a delay in the onset of egg-laying and thus to avoid many small eggs, some large units employ a 23-hour light day for the first 2 months, gradually reducing the light period until it is only 8 hours at 20 weeks. It is then increased to the standard day length for layers of 17 hours.

Battery-cage house

The size will depend upon the system of caging chosen.

1. Vertical tiers, three or four, one above the other, with the droppings tray cleaned mechanically by a scraper or plastic belt.
2. The Californian stepped cage system.
3. The flat deck system over a deep pit.

Because of difficulties over smell, fly nuisance and disposal of droppings, the last system has not found favour. It is generally accepted that houses should not contain more than 25,000 layers.

The most frequent arrangement is to have 3-5 birds in each cage stacked in three tiers. Passageways of 1.0 m are necessary between cages to permit proper ventilation of the bottom cages, and 1.5 m are allowed at the ends.

Floors are concrete over a damp-proof membrane, walls are of various materials according to locality and should have a thermal insulation of 1.7 W per m^2°C. The roof must be carefully insulated (e.g. double skin insulated asbestos, etc.) to give insulation of 1.13 W per m^2°C. If it is a large house it will be necessary to have an egg room attached to one end, unless there is a connecting building joining several houses and an endless belt collecting the eggs as they are laid and roll to the front of each cage, conveying them to a central egg room for cleaning, grading and packing.

The decision of how many birds should be put into a battery cage revolves around what is believed to be a critical space requirement if

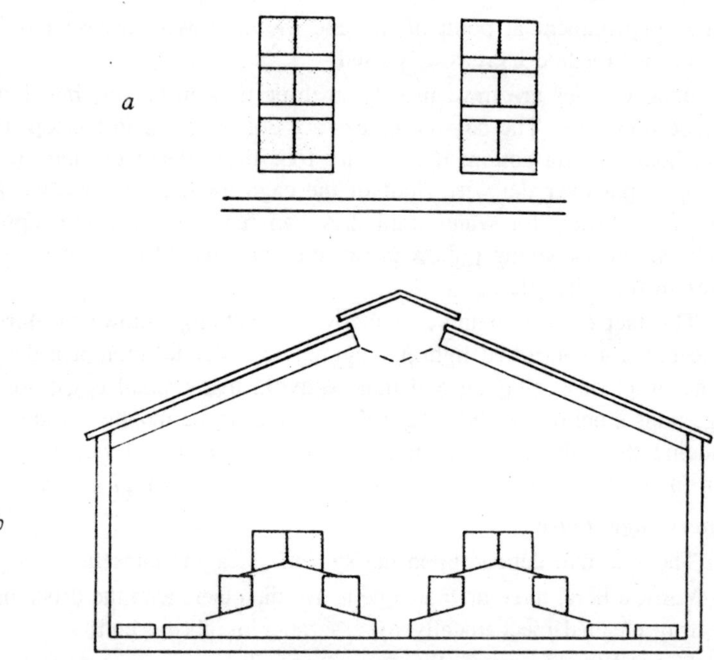

Fig. 2.15. Battery cages for layers. (a) Vertical type; (b) Stepped or Californian type.

egg production is not to suffer or vices such as feather-pecking and cannibalism are not to cause looses. The recommendations of the U.K. Code of Practice, incorporated into official advice, is that in three-bird cages the width should be 500 mm, the depth 430 mm and the height 450 mm. An extra 100 mm in the width is added for each additional bird. It is doubtful whether more than three birds to a cage can be sustained without a varying degree of loss.

Broiler House

The system of producing broilers ready for marketing at 8 weeks is basically similar to the raising of pullets or broiler breeders on deep litter. The difference is that the house is windowless, must be built to have low thermal conductivity and a have a closely controlled ventilation system. This is of importance in the cool but variable climate of many temperate countries. The temperature within the house must be kept as constant as possible at 18°C and can be achieved if there is controlled ridge inlet of air which passes down the smooth internal skin of the roof at 5 m/s (no protruding water pipes or electric cables) is deflected down to the ground by the walls, while the rising heated air is drawn out by wall-mounted fans to the exterior with back-draught shutters. Groups of thermostats placed at a number of places throughout the house will bring on and shut off whatever number of fans are necessary. The ridge inlets are controlled by an electric motor and gearbox operating wire cables. There must be a fail-safe mechanism of having vents and fan apertures opened if power fails. A stand-by generator must be available wherever electricity supply is uncertain.

Watering at 80 mm drinking space per 100 birds and ad lib feeding from continuous lengths of troughing are automatic.

The overall space requirement is 0.9 m^2 per broiler and houses up to 25 m wide can house as many as 20,000 birds. All litter must be cleaned out, and floors, walls and all equipment thoroughly disinfected before giving the house a fortnight's vacancy prior to repopulation.

Tropical Poultry Houses

There has been notable expansion in the number of large units established both for the production of eggs and of broilers in tropical countries during recent years. Most of these have been of open-sided type to take advantage of any cooling breeze. All too often scant attention has been paid to the insulating properties of either roof or walls. Inevitably this has meant that midday temperatures during the hottest months reach levels that result in restriction of appetite, with consequent lowering of egg output or growth in broilers. Even in environment-controlled houses, with low thermal transmission values for roofs and walls, full ventilation rates are not adequate entirely to keep temperatures within the birds' thermal neutral zone all the time.

In many places, care in siting the houses so as to take every advantage of the prevailing wind, a well-insulated roof with overhang and open sides will permit light hybrid layers to reach economical levels of production, given that feeding is optimal and a careful strategy of disease prevention is conscientiously put into effect.

TURKEYS

The traditional small-scale breeding and rearing of turkeys, mainly for the Christmas market, gave way in the post-war years to the large-scale commercial exploitation of much improved new broad-breasted strains from the United States that had faster growth rates and earlier maturity.

Adaptation of the traditional outdoor rearing and fattening methods have characterized many enterprises. The rearing stage has been most intensified with specially constructed buildings housing battery cages, replacing the tiered brooders kept in sheds.

Breeding and Rearing Houses

As the demand for 'oven-ready' turkeys throughout the year has increased, the need for continuous production has encouraged the use of environment-control buildings for both breeding and rearing. The construction of houses for both purposes must follow the requirements already stated for laying hens kept in battery cages.

A similar controlled ventilation system can be used. Lighting must be controlled by a dimming device.

Breeding

Hens and stags may be kept together for natural mating, but there is higher productivity if they are kept in separate cages and artificial insemination undertaken. The space requirements are: 20 kg/m^2 if hens and stags are kept on the floor for natural mating. Stags on their own, whether on the floor or in cages, need 1 m^2, while hens in cages can be kept at the rate of 30 kg/m^2.

Rearing

It is necessary to have a temperature of 21°C during the first few weeks so that heat from hot water pipes or an evenly distributed radiant source must be available. Tiered brooding cages give a higher density per unit of floor space. There must be ready access to food and water at this stage. The lighting should be at a high level.

A temperature of 16°C is adequate for the rearing stage, declining to 10°C at 8 weeks.

Fattening

Until recently the pole barn, constructed as a relatively cheap shelter, or sheds with wire verandahs to give extra space, have been widely used for the varying lengths of fattening period. But with high food costs in relation to the extra energy necessary to offset the low winter temperatures, it has become more profitable to keep all birds in environment-controlled houses. The space requirements increase from 0.14 m^2 for the 8 weeks poult to 0.37 m^2 between 9-16 weeks and 0.46 m^2 for birds over 16 weeks. The length of feeding trough necessary is 7.3 m per 100 birds from 8 weeks, with 50 mm per bird if tubular feeds are used.

DUCKS

The rearing of ducklings for eating has been rapidly increasing, but the demand for eggs is small. Ducks are difficult to rear in intensive systems because of their high water consumption and thus output of liquid excreta. Their webbed feet are also prone to damage when attempts are made to fatten them in battery cages.

The housing requirements for brooding and rearing are similar to those for chickens, for it can be successfully accomplished either in three-tier battery cages holding 25 ducklings up to 3 weeks old or on the floor in divided areas using the same materials for litter as with chickens. Because of the difficulty of keeping bedding even moderately dry, although it is added to each week, many producers have been using wire floors during the first 4 weeks (8 g × 12.5 mm mesh).

For the following 4 weeks until ready for market, various systems are used with a pronounced trend a way from outdoors systems because of the uneconomic food conversion during winter. Ducklings are being fattened on straw bedding, slats and weldmesh (25 mm × 2.1 gauge). Space allowance is 0.18 m^2 per bird at 8 weeks. Access to an outside run with an impervious and well-drained base may be given.

Adult stock need a space allowance of 0.2-0.3 m^2 according to breed. A controlled environment house may be used, divided into slatted areas where are the drinkers and feed trough, with a bedded area behind. Nest boxes are provided in the ratio of one to three ducks.

Rabbits

There is a pronounced trend in many European countries to eat more rabbit meat, with France making the biggest demands. Much better stock and some increase in our understanding of nutritional needs and disease control, have led to the imposition of stricter limits on what to advise about housing. The aim is to have a rabbit weighing 2-2.7 kg live weight which will give a carcass of 0.9-1.4 kg. The does should be able to produce at least 14 litters of 8 young in two years and rear them to slaughter weight in 8 weeks.

This is much more likely to occur if mating, pregnancy, kindling and rearing take place in a controlled environment house. The optimum temperature is 15-16°C and, since it is impossible to crowd breeding rabbits, they are not likely to generate enough heat to avoid much cooler temperatures in the temperate zone winters. There will be some reduction in the efficiency of food conversion, but this may be less costly than installing and running a heating system in the house.

It is of the greatest importance that the ventilation should be adequate. A doe needs 50 m^3 per hour, including the needs of her litter. If growers are housed separately, an allowance of 10 m^3 for each is adequate. Tables are available to show what size of fans and speeds will achieve the right rate. The control of air inlets must be linked to the extraction fans operation in the manner developed by the National Institute of Agricultural Engineering in the U.K. The materials used in the construction of the house and, in particular, the roof, must be highly insulated, the latter having a thermal value of 1.12 W per m^2°C.

The cages now being used are made of welded wire and are placed on concrete stands or suspended by wires over an enclosed space to allow faeces and urine to fall and accumulate beneath it. A convenient height to have the cage is 750 mm from the floor. The minimum area for a doe and her litter is 910 × 61 mm. A stepped arrangement permits a second tier to be stacked over the same floor space but makes access somewhat difficult.

Bucks must be kept in separate cages of at least 0.37 m^2 floor area. Pregnant does must be given a wooden nest box 400 mm long × 300 mm wide × 300 mm high whose top and one side is left open, except for a ledge or retaining board 150 mm from the base to keep the very young rabbits in the nest.

To obtain year-round breeding, it is desirable to provide a regular 17 hours of light each day. This should be of about 10 lux intensity.

3

FEEDING

Many studies have established that the item costing most in obtaining production from farm animals is their food. For all species it is between 70 and 80 per cent. The full production potential of modern strains of pigs, poultry and ruminants can only be obtained by giving them the right amount and the right quality of food. It means that every stockman, like his veterinary adviser, must have an understanding of the principles of animal nutrition.

Farm and domestic animals are divided into monogastrics, those with a single stomach (pigs, dogs, cats, horses, poultry) and ruminants (cattle, sheep, goats). This determines how food is digested and the nature of the substances that are absorbed from the digestive tract.

The food given to an animal or that it obtains from grazing has four main functions:

1. It provides *energy*. The constituents in the ration that contain energy are the various classes of carbohydrates, proteins and fats. They are broken down by the processes of digestion into simple substances, absorbed and distributed to the tissues. After undergoing many changes within the cells, energy is utilized to maintain life processes and frequently for a large number of production purposes, growth, reproduction, lactation including storage as fat.
2. It provides *proteins*, which are substances containing amino (NH_2) nitrogen essential for muscle growth, wool, milk and egg production. Only protein surplus to the animal's need is used for energy.
3. It provides *minerals*, used in bone formation and maintenance, and in a large number of vital enzyme systems.

4. It provides *vitamins*, also important in enzyme systems, some associated with cell permeability, others with cell intermediary metabolism.
5. It provides a proportion, at least, of the *water* that all animals need. The fact of the varying amount of water that feedingstuffs possess makes it necessary to base all comparisons of the feeding value of what the animal eats on its *dry matter*. This is simply what is left after all the water is driven from it.

Energy

There must be a unit of measurement of energy. The energy within the food and its final utilization within the body can be measured as heat. The unit is therefore the joule. This has replaced the calorie, the relationship being that 1 calorie = 4.184 joules.

Examples of the energy content of well-known chemicals that are to be found in feedingstuffs and of some common foods themselves are given in the following list.

Food constituents:		*Amount kJ/g DM*
1. Carbohydrates	Starch	17.16
	Glucose	13.17
	Cellulose	17.56
2. Protein	Casein	23.44
3. Fat	Butter fat	38.68
	Oil (from seeds)	39.18
Foods:	Barley	18.3
	Maize	19.0
	Barley straw	18.0
	Grass hay	17.7
	Soya bean meal	19.5

It will be seen that fats have approximately 2.25 times as much energy per unit mass. Thus not only when fat itself is added to a diet, as it often is in the feeding of calves and broilers, but also when supplements of oil-seed meals (soya-bean meal, groundnut meal, etc.), are used, they contribute not only protein but fat as well.

Proteins

Young animals of all species require in their diets the building blocks for muscle growth, the amino acids. These consists of a nitrogen-containing amino group (NH_2) and a non-nitrogen or carboxyl group

(COOH). Proteins are made up of these units, some 25 of them, divided into essential and non-essential amino acids. The difference is that the former cannot be synthesized in the tissues and have to be provided in the diet given to monogastric animals. Ruminants are able to depend upon the microorganisms within their rumens to synthesize the essential amino acids, although they commonly receive them in their feed also. As the microorganisms are digested lower down the intestinal tract, the amino acids are absorbed.

If protein in a ruminant's feed is in small amounts, it is necessary for there to be an intake of nitrogen in some other form to provide enough for the optimal growth of the microorganisms within the reticulo-rumen. There are, in the plants that make up pastures, varying quantities of non-protein nitrogen in the form of amino acids themselves, nitrates, amides and amines that help to fill this need. A common source of non-protein nitrogen used in making up rations for ruminants is urea. It has to be borne in mind that some of the essential amino acids (methionine, cystine, cysteine) have sulphur in their make-up, so that this element must be in sufficient amount in the food provided; in most circumstances it is.

Minerals

From the quantities of the minerals found in the body, and thus what has to be supplied in the feed, one can see that there is a division between those needed in relatively large amounts, the major minerals, and those needed in very small amounts, the minor or trace elements.

Table 3.1. Essential mineral elements and their approximate concentration in the animal

Major element	*Approx. concn, %*	*Minor element*	*Approx. concn, mg/kg*
Calcium	1.5	Iron	20-80
Phosphorus	1.0	Zinc	10-50
Potassium	0.2	Copper	1-5
Sodium	0.16	Manganese	0.2-0.5
Chlorine	0.11	Iodine	0.3-0.6
Sulphur	0.15	Cobalt	0.02-0.1
Magnesium	0.04	Molybdenum	1-4
		Selenium	1.7
		Chromium	0.08

Table 3.1. shows those elements that are at present known to be essential for an animal's or bird's health. There are many more minerals found in varying amounts in the food eaten and it is possible that in the future some of these will be found to play key roles in metabolic processes within the cell.

The major minerals, calcium, phosphorus and magnesium are components of the bone-salt, so that a large part of their total concentration is involved in this tissue. Nevertheless, an important proportion is to be found in the fluids and cells of the body, along with all the other essential elements, performing vital functions as parts of enzyme systems or maintaining the acid-base balance and other electrochemical roles.

It has also to be remembered that some of the minerals can be toxic if given in excessive amounts. Instances of this are the effect of too much molybdenum, selenium or copper that occurs in some pastures, either through there being 'accumulator' plants present or too high a concentration of the mineral in the soil.

Vitamins

These are organic compounds that are required by plants and animals in small amounts for normal maintenance and production. They are conveniently divided into those that are *fat soluble*: retinol (vitamin A), cholecalciferol and ergocalciferol (vitamins D_3 and D_2), tocopherols (vitamin E), phylloquinone (vitamin K_1); and those that are *water soluble*: thiamin (vitamin B_1), riboflavin (vitamin B_2), nicotinamide, pyridoxine, pantothenic acid, biotin, folacin, choline, cyanocobalamin (vitamin B_{12}), ascorbic acid (vitamin C).

Many of the water-soluble vitamins are synthesized by bacteria in the intestine of monogastric animals and all of them by the microorganisms of ruminants and herbivores like the horse.

The most important fat-soluble vitamins, for example retinol and cholecalciferol (vitamins A and D_3), have their precursors or pro-vitamins in plants, but the greatest concentrations of the preformed vitamins are to be found in animal sources, particularly in fish-liver oils.

Provitamins

The provitamins of retinol (vitamin A) are the carotenoids, found in growing plants and associated with the chlorophyll or green-colouring pigment. The most widely distributed and most active of them is β-carotene, which is converted to vitamin A in the cells lining the gut and in the liver. This association with the greenness of forage plants gives a good indication of the adequacy of the diet of cattle, sheep and horses as sources of retinol. The conversion is approximately 3 μg β-carotene = 1μg retinol. With this vitamin, the liver has the capacity to store several months' need.

The source of vitamin D_2 for grazing animals is a phytosteroid, ergosterol, found in the leaves of pasture plants. Following irradiation from the ultra-violet region of sunlight, it is changed to ergocalciferol. Another source of this vitamin is a sterol in the skin, 7-dehydro-cholesterol, which is converted in the same manner to cholecalciferol (vitamin D_3). Both forms of the vitamin have equal biological value for all farm animals, except poultry, where vitamin D_3 is many times more active. This vitamin may be stored in fat depots of the body to carry young animals over a limited period of deprivation, such as occurs in the winter months in northern latitudes when ultra-violet radiation is deficient.

Antivitamins

The water-soluble vitamins are not stored, but some may be destroyed within the digestive tract by specific enzymes and thus render the animal deficient.

This may occur with *thiamin* (vitamin B_1). It has been shown that certain strains of bacteria, occasionally able to grow within the rumen of sheep and calves, produce an enzyme, thiaminase, that destroys the vitamin and signs of a severe deficiency may follow, the disease being known as cerebro-cortical necrosis. Bracken fern, if eaten in large amounts by horses, and raw fish, sometimes fed to silver foxes and mink, have similar thiaminases and can thus cause illness through thiamin deficiency.

Analogues, chemical compounds with very similar structures to various water-soluble vitamins, but with the power of inactivating them, have been found in extracts of natural substances or been synthesized, but most are of no practical importance.

The only fat-soluble antivitamin of marginal practical interest is dicoumarol, found in spoilt sweet clover (*Melilotus alubs*). It destroys the vitamin's function as a co-factor in the formation of prothrombin, essential for blood clotting.

DIGESTION

Monogastric Animals: Pig, Dog, Cat, Horse

Food eaten by monogastric animals is mixed with saliva which, from its mucin content produced by the submaxillary glands, has a lubricating effect, thus assisting swallowing. There is a digestive enzyme, ptyalin, in the saliva, which can begin the breakdown of sugars and starch. Dogs, cats, and pigs swallow their food quickly, so there is little digestion until the food has been in the stomach for a while. Here, high acidity combined with pepsin effectively breaks the peptide bonds holding together the polypeptides of the food proteins.

The small intestine receives digestive enzymes from the pancreas and from one of the types of epithelial cells that make up its lining. Bile comes from the liver through its own duct. The effect of the enzymes brought along in these secretions and the emulsifying action of bile salts on fats along with its activation of the fat-splitting enzyme lipase, result in the final conversion of the major dietary constituents of carbohydrates, proteins and fats into their component parts of glucose, amino acids and fatty acids. These can be absorbed into the body tissues.

The cells of the large intestine do not produce digestive enzymes but there is a lot of microbial fermentation of food residues. In the pig there is breakdown of about half of the relatively small amount of cellulose in the digesta, but it is otherwise in the horse. The caecum and anterior colon of the horse are of great size and have a large and varied population of bacteria and protozoa. These breakdown cellulose

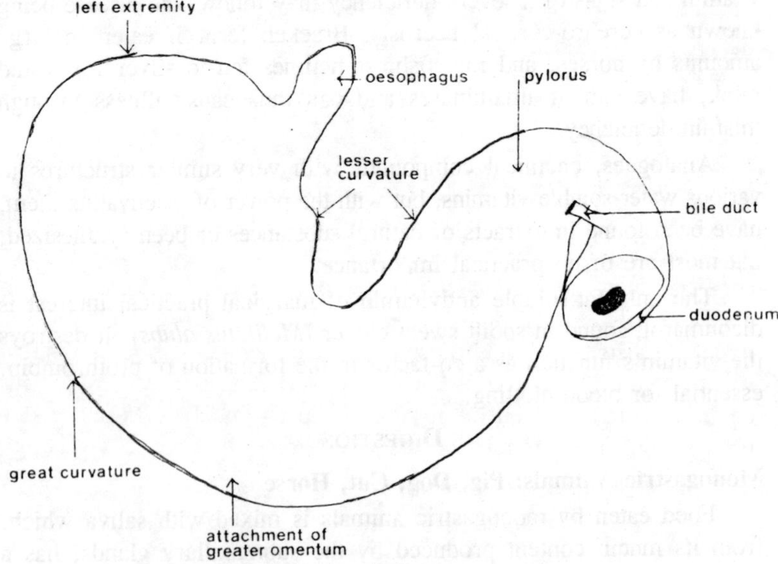

Fig. 3.1. Diagrammatic form of a monogastric stomach.

and hemicellulose to the same simple organic acids (acetic, propionic and butyric acids) that occur in the fore-stomach of ruminants. There is now evidence that some of the microbial protein is absorbed from the horse's colon.

Fowl

Food is picked up by poultry and immediately swallowed, to enter first the diverticulum or crop, a storage compartment in which a degree of fermentation occurs. The food then passes into the proventriculus or glandular stomach, where the finer particles of food come under enzyme attack. Beyond it is the gizzard, a very muscular organ with rigid toughened lining, where the grinding of any larger pieces of food occurs and its mixing with water to produce a paste. In the small intestine, the enzymatic reduction of the digesta proceeds in the same way as in mammals, along with the absorption of nutrients.

Ruminants

The remarkable and important difference in the digestion of ruminants is the size and nature of the four compartments of the stomach. The first, the rumen, is the largest, having attached to it an opening into the much smaller reticulum in a forward position. Also opening from the rumen, lying to the right of it, is the omasum, which is small and round and followed by the true stomach, the abomasum. The disposition and functioning of the rest of the intestinal tract are similar to the simple-stomached animals.

The rumen

The purpose of the reticulo-rumen must be understood if the feeding of ruminants is to be effectively and economically accomplished. It is virtually a fermentation device in continual motion, mixing the contents. Some of the larger pieces of forage, as they come close to the cardia, or entrance to the rumen, pass through it and by a wave or reverse peristalsis, are brought back into the mouth for a second more thorough chewing, known as *rumination*. As the food in the reticulo-rumen attains a degree of fineness and liquidity, as a result of breakdown by microorganisms, it is moved on to the true stomach. Large amounts of saliva, some 50-60 litres being the daily amount secreted by a steer, containing urea and buffering salts, help to liquefy the food and to provide the microorganisms with a source of nitrogen as well as preventing the rumen contents becoming acid.

Fermentation results from the activity of various species of bacteria and protozoa. The rate of breakdown of the food within the rumen, as

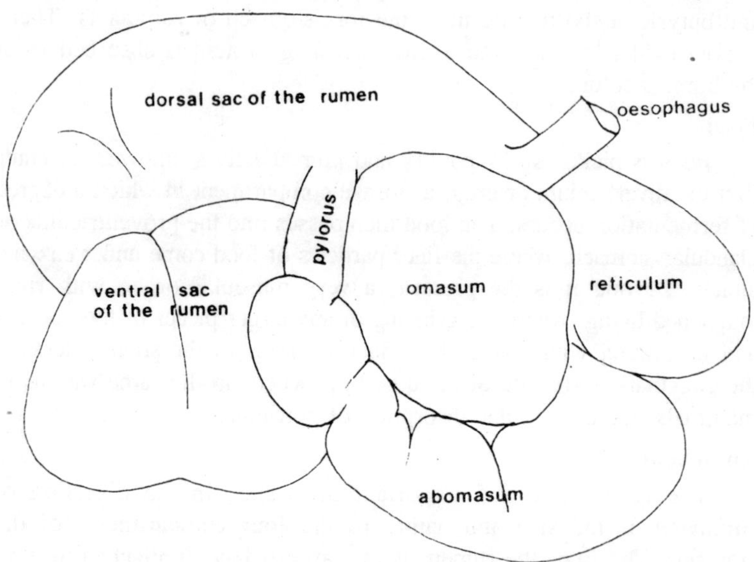

Fig. 3.2. Ruminant stomach.

well as the amount and exact nature of the product, depend upon what is fed. For example, if a ewe or cow is grazing young spring pasture, the rate of fermentation of cellulose and hemicellulose will be faster than if hay from the same pasture is fed. The reason is the much higher proportion of sugar, amino acids and other relatively simple chemical compounds in young pasture that act as suitable substrates for a rich variety of microorganisms.

Products of fermentation

Carbohydrates

The products of this fermentation are a mixture of mainly lower volatile fatty acids, acetic, propionic and butyric acids (C_2, C_3, C_4 compounds), and ammonia (NH_3) from the degradable nitrogenous substances in the food. The organic acids are absorbed through the rumen wall to pass in the portal circulation to the liver, where the propionic acid is used to produce glucose and much of the acetic and β-hydroxybutyric acids go directly via the bloodstream to various organs and tissues, where they are used as sources of energy and fatty acids.

Degradable and non-degradable protein

The ammonia nitrogen is utilized along with the available carbon skeletons, plus energy and minerals, particularly sulphur and phosphorus, to build up a quantity of bacterial and protozoal protein which passes into the rest of the tract to be digested and absorbed. At the same time, if there is sufficient protein in the diet to result in a surplus in the rumen, this proportion will not be degraded and so will pass to the small intestine for digestion. Thus there is rumen degradable and non-degradable protein.

Any non-protein nitrogen, which will include any urea given in the diet, will be degraded first. The more soluble proteins, such as casein in milk, will be degraded before less-soluble food proteins, such as the protein of maize. The latter stand a better chance of avoiding deamination by the rumen microorganisms.

The measure that has been long used to state the protein needs of livestock has been 'digestible crude protein' (DCP), but its use has become increasingly suspect because it gives no indication of how much of it escapes rumen degradation to pass onward and be digested and absorbed from the small intestine. It is this part of the dietary protein that is so important as the source of amino acids for purposes of production, that is growth in the young animal, milk and wool production.

The close relation of available energy and the extent of microbial degradation of protein in the diet has been more closely examined in the past few years. If there is a sufficient supply of energy, water, minerals and bicarbonate from saliva, the micro-organisms will synthesize approximately 30 g of microbial protein nitrogen per kg of organic matter digested (DOM).

It is also known that the proportion of DOM that is digested in the rumen is between 0.6 and 0.7, so the figure of 0.65 has been taken to calculate the likely relation between the amount of rumen degradable protein (RDP) and the energy expressed as metabolizable energy (ME). This is found to be;

$$\text{RDP (g per day)} = 7.8\ \text{ME}$$

The practical implication of this relationship has been seen in the way that poor quality forages, such as straw, having little protein and digestible energy, can only be marginally improved by feeding urea with them. But if the straw is treated with alkali to increase its digestibility, the added nitrogen from urea can than be degraded and used in the multiplication of microorganisms. Other low protein but high energy feeding stuffs, such as maize silage and barley, are eaten in greater amounts and more completely digested when supplemented with urea.

A further development has been the calculation of the contribution that the RDP makes in the form of tissue microbial protein (TMP) to the total tissue protein (TP) need to the animal. This latter includes the protein needed, not only for maintenance, but also for growth, for the protein content of milk, for wool growth and for the growth of the foetus and associated tissues during pregnancy.

The TMP, or microbial protein, is usually sufficient for maintenance, plus the growth of cattle above 230 kg live-weight, or a modest milk yield of 7.5 kg daily. Production above these levels depends upon the quality and quantity of undegraded dietary protein (UDP).

This is related to the total tissue protein requirement by the equation:

$$\text{UDP (g per day)} = 1.91\ \text{TP} - 6.25\ \text{ME}$$

A measure of degradability

The total protein requirement for any ruminant is thus the sum of RDP and UDP. To make this calculation, it is necessary to know the extent of the *degradability* of the proteins found in the common forages and concentrate ingredients currently fed to ruminants. At the moment many of these have not been accurately assessed; but for the time being, they are grouped in the following manner:

Case	*Range, dg*	*Forages*	*Cereal*	*Protein supplement*
A	0.71-0.90	Grass hay Legumes hay Grass silage (untreated) Artificially dried grass	Barley Wheat	Groundnut meal Soya bean meal (untreated) Sunflower meal Rape seed meal
B	0.51-0.70	Grass (fresh) Legumes (fresh) Artificially dried grass (pelleted) Maize silage	Maize	Soya bean meal (treated) Linseed cake Fishmeal
C	0.31-0.50	Artificial dried legume (pelleted)	Milo	Herring meal Fish and casein
D	0.31	Grass silage (formalin treated)	—	—

Even when given a satisfactory energy intake (11 MJ per kg DM per day), a young steer up to at least 200 kg finds RDP insufficient

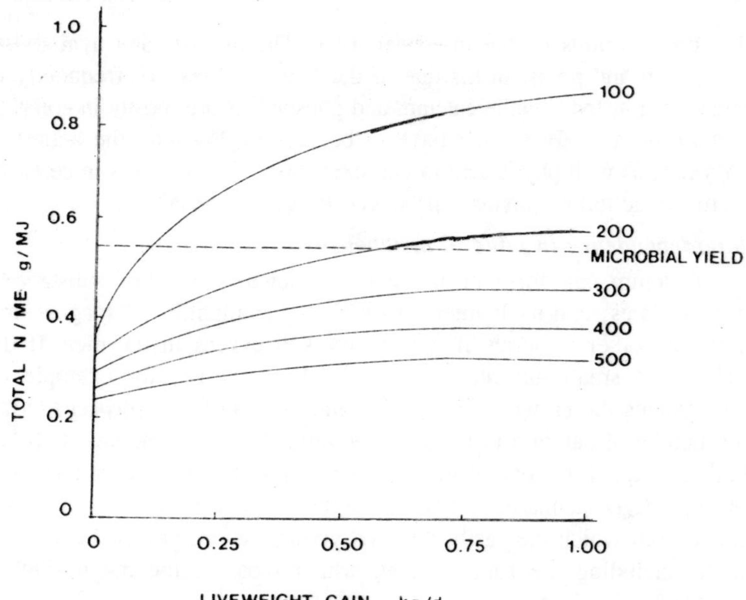

Fig. 3.3. Relationship between live-weight gain of steers at various live-weights given a diet of 11 MJ per kg to meet energy needs and tissue nitrogen requirement per MJ.

for rapid growth, although enough for the heavier steer. As additional knowledge of the variation in the amount of endogenous or unavoidable nitrogen loss on different diets is obtained, these levels will vary.

The significance of this development in the feeding of ruminants is that it focuses attention on the need to find new ways of reducing the degradability of at least a part of the protein fed to high production animals. At present, cows are sometimes given too much rumen degradable protein (RDP) with consequent wastage as ammonia.

Minerals

Minerals are not digested as are organic materials of the diet. They are absorbed from various parts of the intestinal tract and at widely varying proportions of the amount obtained in the diet. While sodium and potassium are absorbed almost completely, some trace elements, particularly copper and zinc, are only absorbed to the extent of 10 per cent.

There are many factors influencing the availability of minerals, the more important being the form in which the mineral occurs in food, the overall rate of digestion and the variation in the pH of different sections of the intestinal tract. The almost total availability of sodium and potassium is due to the fact that they are frequently in ionic form in food, while calcium and phosphorus are mostly in complex metallo-organic forms that have to be broken down by the action of enzymes, as with phytic acid (a common form of phosphorus in cereals), from the action of phytase, produced by gut bacteria.

Interdependence in mineral metabolism

Calcium can form insoluble components with other substances, e.g. oxalates, and itself interferes with the availability of magnesium, zinc and other minerals if it is in marked excess in the diet. If the pH of the small intestine can be slightly lowered, for example by encouraging the growth of non-pathogenic strains of *Streptococcus lactis*, the uptake of calcium in monogastric animals can be increased. It has been suggested that one of the factors responsible for the high incidence of hypomagnesaemia in dairy cattle in the spring is the greatly increased speed with which the easily digested young pasture passes through the tract, including the rumen, from which most of the magnesium is absorbed.

Important practical deficiency states in grazing sheep and cattle occur when there is a higher than normal intake of molybdenum and sulphate for these will prevent the uptake and liver storage of copper under conditions where the intake of the latter element might otherwise have been just sufficient.

A concept that was developed by workers at the Institute of Research in Animal Diseases in the United Kingdom a few years ago, is valuable in considering the minerals that are of particular importance in the production of metabolic disease. It consists of considering the dynamic balance, as far as there is information available to do so. It involves knowing the input of the mineral being considered, the extent and availability of whatever body reserves there are and the output of that element.

With so small a part of the body reserves of magnesium and calcium available to be mobilized if the intake becomes inadequate for the production requirements, it is easy to see how a deficiency state occurs.

The absorption capacity of the intestinal tract for some major minerals, particularly calcium, diminishes with age. A calf will absorb about 55 per cent of dietary calcium; but by the time it is 12 months old this will have fallen to 45 per cent and in an old cow will be even less. Numerous studies in mineral metabolism have emphasized the fact that there is large individual variation in the manner in which animals of all species deal with mineral in their diet. In the recommendations for levels of intake of the essential minerals, an approximate margin is allowed to safeguard this inevitable variation.

Feeding Horses

Most of the research upon which the feeding standard for horses is based has been undertaken in the U.S.A. and therefore has particular relevance to the feedingstuffs commonly used in that country. Wherever horses are used, it is clear that the need for energy will directly depend upon the amount of work undertaken. Some trails at Cornell showed that the energy demand of ponies playing polo was over 20 times that of the resting animal.

It is convenient to consider the requirements for both energy and protein in two parts: (1) when the animal is at rest and neither gaining nor losing weight, i.e. its *maintenance* needs; (2) when it is at work, which should be defined, if possible, in terms of the *time* and the *nature* of the activity. What we know as the condition or degree of training, the ability and weight of a rider, all make a difference to how much additional energy must be given above maintenance. Horses, especially race horses, exhibit notable individual variation in their efficiency of food utilization, so that it is important to adjust, as a result of daily observation of whatever extent is necessary, the levels of intake stated.

Table 3.2. Tables of daily requirements

Weight, kg	*Age*	*Digestible energy, MJ*	*Digestible protein, g*	*Dry matter intake, kg*
200	Adult	34.6	140	3.15
170	Yearling	34.0	200	2.90
400	Adult	59.2	240	6.30
265	Yearling	57.2	350	4.95
500	Adult	68.7	290	7.45
300	Yearling	70.6	450	6.00

Energy

The horse, like other monogastric animals, depends upon carbohydrates as a main source of energy, although fats and protein may also be used. If they are allowed to choose, horses will select food with a high concentration of digestible energy, that is, they will select grain rather than a forage and high quality rather than poor quality hay. The measure used is the amount of energy digested—the *digestible energy* (DE). This is the total energy contained in the food, the gross energy less the energy in the faeces voided by the horse.

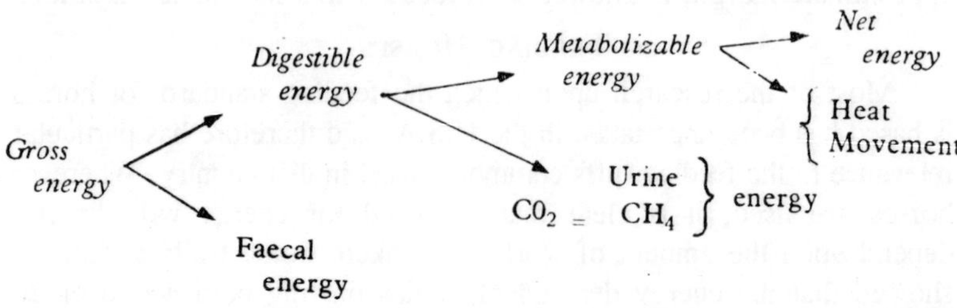

Not all the digestible energy is available for use by the horse, for there are small and varying losses in the urine, in gases generated as a result of fermentation and a loss from the metabolic processes of finally making the energy available for use in the cells. This is called the *net energy*, but lack of quantitative information about the size of losses in the various stages of digestion of feedingstuffs given to horses makes it impossible accurately to calculate net energy at present.

Maintenance

With horses, as with all animals, there is a direct relationship between the metabolic weight and the energy needed for maintenance. For the horse this is

$$\text{DE (kcal/day)} = 155\ W^{0.75}\ (W = \text{wt in kg}).$$

Table 3.3. Values for metabolic weights at 50-kg intervals to 590 kg

Weight of horse, kg	*Metabolic weight,* $W^{0.75}$*, kg*
100	31.6
150	42.6
200	53.2
250	62.9
300	72.1
350	80.9
400	89.4
450	97.7
500	105.7
550	113.6
590	119.7

Production

The allowances necessary for the work done by riding horses, to be given in addition to maintenance, have been found by American workers to be as follows:

Activity	*Digestible energy per hour*	
	kcal	*kJ*
Walking	0.5	2.1
Slow trotting, some cantering	5.0	20.9
Fast trotting, cantering, some jumping	12.5	50.2
Cantering, galloping, jumping	23.0	96.6
Strenuous effort (polo), racing at full speed	39.0	163.8

Pregnancy

The pregnancy of the mare is approximately 334 days, but the main growth of the foetus takes place during the last 90 days. A mare in good condition at the time of conception can safely be given a maintenance ration. During the critical three months at the end of gestation, the energy intake should be increased by 12 per cent.

Lactation

The energy content of mare's milk is 2 MJ/kg. It is believed that she converts digestible energy with approximately 60 per cent efficiency, so that 3.3 MJ of DE are needed for each kilogram of milk. The mares of light breeds can produce 24 kg milk/day at peak lactation which occurs at about two months after foaling, but the mean is 12-18 kg. It is suggested that an extra energy allowance of 3 per cent of body weight be made during weeks 1-12, reducing to 2 per cent for the latter 13-24 weeks of lactation.

Growth

All young animals have the potential of fast growth both while they are suckling and after weaning. A pony likely to weigh 200 kg when adult, at the age of 6 months will gain 0.5 kg/day and at one year its rate of gain will have fallen to 0.2 kg/day. A horse that will become 500 kg when fully grown will gain weight at 0.8 kg/day at 6 months and about 0.55 kg/day as a yearling. The present advice is to provide energy at close to the same level that the young horse will need for maintenance when it attains its adult weight.

Protein

The horse, as a simple-stomached animal, digests proteins with moderate efficiency in the fore-part of the intestinal tract and, through the synthesis and partial absorption of microbial protein in the lower tract, digests a higher proportion of nitrogenous food ingredients than ruminants.

Maintenance

The amount of digestible protein (DP) needed has been related to the biological weight as 27 g/$W^{0.75}$(kg). Good quality grass hay provides an adequate balance of amino acids for maintenance. For a pony of 200 kg, 140 g DP/day, for a 400 kg horse, 240 g DP/day and for a 500 kg horse, 630 g DP/day are adequate.

Production

Growth

Present evidence suggests that after the maintenance requirements for protein have been satisfied, 45 per cent of the remaining DP is used for growth. This means that a pony 6 months old will need over twice as much protein as the adult is given for maintenance, i.e. 310 g DP/day, reducing to 200 g DP/day at one year. A horse whose final weight will be 500 kg, when 6 months old will require 520 g DP/day and 450 g DP/day as a yearling.

Work

It is an interesting fact that a horse in work does not require additional protein. A little protein is lost in the sweat, but this is made up from the additional food needed for energy.

Pregnancy and lactation

It is not until the final three months of pregnancy that the demand for extra digestible protein must be met. Approximately 100 g DP/day is sufficient and, with the normal DP content of cereals (e.g. oats) at least 9 per cent, the extra grain (600 g) being provided will give the mare over half the extra protein as well. If an oilseed meal, such as soya-bean meal, is added to the cereal at 15 per cent, the extra protein will be provided

In lactation the protein content of mare's milk drops from 3.1 to 2.2 per cent within two months and it has been found that the larger quantity of the above concentrate mixture fed to a stabled mare to satisfy her energy demand will also provide the additional protein.

Minerals

It is necessary to be careful to maintain a ratio of approximately 1:1 for Ca:P, especially with young horses. Ratios of 2-3 Ca: 1 P can be sustained safely in older horses. The absorption of these minerals from the intestinal tract varies with age, but is 55-75 per cent for calcium and 35-55 per cent for phosphorus. The requirement for a hunter of 500 kg averages 20 g calcium/day and 12 g phosphorus/day. This would have to be increased for a brood mare of this size by 6 g Ca and 3 g P during the last 90 days of pregnancy, and for lactation 15 g Ca and 7 g P.

The only other minerals that must be provided additionally to those contained within the diet are sodium and chlorine (as chloride) because of the loss through sweating in worked horses. The requirement is satisfied if salt is given at 0.1-0.5 per cent of the diet. In practice, a piece of rock salt, or a salt lick is placed in the stable or loose box.

Vitamins

Of all species of farm and domestic animals, the horse is able to obtain sufficient of all the vitamins without additions to the diet, so long as it is given good quality hay, made from mixed pastures or legumes if stabled, or is able to graze fresh green forage for at least 6 weeks. This latter is sufficient to provide enough retinol (vitamin A) for 3-6 months, even when there is very little provitamin or the vitamin itself in the diet. It will obtain sufficient vitamin D from exposure to sunlight, and the water-soluble vitamins are synthesized in the gut as well as being contained in feedingstuffs.

The most accurate rationing depends upon knowing an analysis of the feedingstuffs available. If such estimates are not available, one has to depend upon the average values shown in tables.

If a pony weighing 200 kg is taken as an example and average values for good quality mixed grass hay and oats are taken, the amount to be fed can be calculated as follows:

	DM, g/kg	*DE, MJ/kg*	*DCP, %*
Good quality grass hay	90	7.15	5.5
Oats	89	14.03	10.5

With the need for approximately 35 MJ of DE per day, it is clear that if the pony is given hay alone at 5 kg it will satisfy the energy needs. A hunter weighing 400-500 kg must be fed according to the amount of work expected. On rest days the quantity of concentrate food must be reduced.

The feeding of *racehorses* in practice tends to vary widely according to predilections of trainers. In order to encourage maximum energy intake in highly strung individual animals, various unusual items may be included, such as carrots, dried milk powder, as well as flaked maize and bran. The need for a reduction in roughage intake and an increase in concentrated highly digestible feedingstuffs in the days preceding a race is greatest for steeplechasers and distance horses.

Table 3.4. Practical daily feeding regime

	Good quality hay, kg	*Cereal (rolled oats, rolled barley, cracked wheat or maize), kg*	*Protein supplement (soya-bean meal, linseed meal), kg*	*Minerals, g*
Rest days	8	1.5	0.3	60
Light work	7	3.3	0.5	60
Heavy work	5	6.6	0.5	60

DAIRY COWS

Energy

An important difference in the manner in which ruminants digest their food compared to monogastric animals is the loss of energy from the methane gas formed as a result of fermentation in the reticulo-rumen. It represents about 0.08 of the energy value of the food at maintenance, falling to about 0.06 at higher levels of intake. When this loss and the much smaller loss of energy in urine is subtracted from the digestible energy, we are left with the *metabolizable energy* (ME) value of the food, and this is measure used in rationing ruminants. About 0.81 of the digestible energy is utilized or metabolized by the animal.

Many trails have been conducted with both sheep and cattle to discover the digestibility of the main components of feedingstuffs, the carbohydrates, measured as crude fibre (CF), nitrogen-free extractives (NFE), crude protein (CP) and fat, measured as ether extract (EE). From these studies metabolizable energy can be calculated by the following formula:

$$\text{Metabolizable energy (MJ/kg)} = 0.0152\ \text{DCP} + 0.0342\ \text{DEE} + 0.0128\ \text{DCF} + 0.0159\ \text{DNFE}$$

Maintenance

Although the most accurate relationship is with the metabolic weight of an animal, it has been agreed that for practical purposes direct relation to the animal's weight can be used.

$$\text{Maintenance (MJ/head)} = 8.3 + 0.091\ \text{W}.$$

Table 3.5. Daily maintenance allowance for cattle

Weight, kg	*Allowance, MJ/head*
100	17
150	22
200	27
250	31
300	36
350	40
400	45
450	49
500	54
550	59
600	63

Production

Milk

The energy allowance for milk production must be based upon the energy in the milk, itself dependent upon the amount of fat (BF) and solids-not-fat (SNF) that it contains. This can be obtained from the relationship:

Energy value of the milk secreted = 0.0386 BF + 0.0205 SNF –0.236

and since the utilization of metabolizable energy for milk production is accomplished at 0.62, the extra energy that must be fed for every kilogram of milk produced can be calculated. For what is known as 'solids corrected milk' (SCM) in which milk is converted to a standard of 4 per cent fat and 8.9 per cent solids-not-fat, it is necessary to feed 5.31 *MJ/kg milk*. With milk of average composition where most of the cows are of Friesian type, with 3.6 per cent fat and 8.6 per cent solids-not-fat, it is 4.94 *MJ/kg*.

Weight change

Whether it is during the first six weeks after calving when heavy milkers inevitably lose weight, or as a result of shortage of food, the cow will call upon her body reserves to make up the deficit. Body tissue has an energy value of 20 MJ/kg. It is believed to be used with 0.82 efficiency for milk production, so that each kilogram of body energy mobilized by her permits the secretion of 16.4 MJ as milk (20 × 0.82). This, with an agreed safety margin, will be equivalent to (16.4 × 1.05)/0.62 = 28 MJ of dietary ME per kg weight lost.

Similarly, if the supply of energy is more than is necessary for the production of milk, the cow will gain weight. This is the normal state of affairs by about the 120th day after calving, probably at the rate of 0.5 kg/day. In the lactating cow this has an efficiency of 0.62, so that it will be necessary to make an allowance of (20/0.62) × 1.05 = 34 MJ/kg of weight gained.

Pregnancy

It has been calculated that the requirement of extra energy for pregnancy is only 5 MJ/h/day until the sixth month. During the last three months there is need for an extra 70-80 MJ/h/day.

Dry Matter Intake

A main factor controlling the amount of food eaten by an animal is its size; but others are the amount of milk a cow is producing, the digestibility of the ration and the manner in which forage has been presented, i.e. a silage with a short period of fermentation is more palatable than one with a higher pH value due to longer fermentation. During the first two months of lactation, a cow's appetite is reduced so that more concentrated foods must be fed. A guide to the dry matter intake (DMI) of cows at different stages of lactation and giving varying yields.

Proteins

Until recently the requirements for protein for dairy cows have been simply expressed as digestible crude protein (DCP) as follows:

		Digestible crude protein for	
Breed	*Weight, kg*	*Maintenance, g/day*	*Production g/kg milk*
Charolais	350	230	66
	400	250	
Ayrshire	500	290	56
Friesian	600	330	51
	650	350	

To permit a greater precision of rationing, it is now necessary to consider the protein requirements in two parts, the rumen degradable (RDP) and the undegraded dietary protein (UDP). Tables are shortly to be issued which show the requirement of both for cattle and sheep of differing weights and showing the animals at maintenance or having various levels of production.

Example:

A Friesian cow weighing 600 kg and giving 30 kg of 36.8 g fat/kg milk. The feed available is grass hay, rolled barley and soya-bean meal whose food values are:

	DM, *g/kg*	*ME,* *MJ/kg* DM	*RDP,* *g/kg*	*UDP,* *g/kg*	*Degradability,* *dg*
Grass hay	850	8.4	68	17	0.80
Rolled barley	860	13.7	86	22	0.80
Soya-bean meal	900	12.3	302	201	0.60

The first step is to look up a table to discover how much dry matter the cow is likely to eat; then state its maintenance and add to this the production allowances.

Dry matter intake (DMI) = 18 kg/day

$$\left.\begin{array}{rl} \text{Maintenance} & = 63 \text{ MJ of ME/day} \\ \text{Production} & = 139 \text{ MJ of ME/day} \end{array}\right\} 202 \text{ MJ/day}$$

	DMI, *kg*	*ME,* *MJ*	*RDP,* *g/day*	*UDP,* *g/day*	*Degradability* *dg*
Total requirement	18	202	1575	597	0.72
12 kg grass hay	10.2	85.7	694	173	0.80
8.5 kg rolled barley	7.3	105.5	662	169	0.80
1.0 kg soya-bean meal	0.9	11.1	272	180	0.60
	18.4	202.3	1628	522	0.72

This ration, while satisfying the requirements for energy and within the appetite limit, provides an unbalanced protein intake, with an excess of rumen degradable protein and a deficiency of undegradable protein. The situation can be put right if the barley is reduced and some flaked maize added. The latter has a lower protein degradability, presumably because of the heat treatment during processing reducing the solubility of zein (the protein of maize). Keeping the forage intake constant and giving 4.5 kg rolled barley with 4 kg flaked maize and 0.75 kg soya-bean meal daily preserves the limits of DMI and ME, and the small deficiency of RDP of 108 g could be made up by adding 43 g/day of urea since urea has 46 per cent nitrogen. This is calculated as follows:

$$\frac{\text{Deficit of RDP}}{6.25 \times 0.80 \times 0.46} = \frac{108}{23} = 43\text{g per day.}$$

Taking the same cow, but using grass silage and a dairy cake as balancer, it has been shown by Webster (1980) that, again, it is by using less degradable forms of protein that high production can be sustained.

Friesian giving 30 kg milk/day						
	DMI, kg/day	*ME, MJ/day*	*Crude Protein, g/day*	*RDP, g/day*	*UDP, g/day*	*dg*
Total requirement	18	202	2175	1575	600	0.72
Intake						
40 kg grass silage (250 g/kg DM, 10.1 MJ/kg DM)	10	102	1500	1200	300	0.80
Either:						
9.5-kg cake (860 g/kg DM, 12.5 MJ/kg,	8	100	1084	1028	256	0.80
6% CP, dg 0.8)	18	204	2584	2228	556	
Balance	0	0		+768	-44	
Or:						
9.5-kg cake (860 g/kg DM, 12.5 MJ/kg, dg 0.55)	8	100	675	375	300	0.55
	18	204	2175			
Balance	0	0				

With the first type of dairy cake the protein is being used inefficiently because of the excess degradable protein in the rumen, while the undegraded protein is not sufficient for milk production. The need therefore is to make a cake of low protein degradability (0.55) with normal metabolizable energy constant, for the actual protein requirement has been almost halved (675 g instead of 1280 g). In this way excess degradable protein, with wasteful extra ammonia production occurring in the rumen, is avoided and useful additional amino acids are available in the small intestine for milk production.

Developments in Practical Feeding of Dairy Cows

There have been two interesting developments in the way in which cows are fed during the 180 days of indoor feeding that are conventionally taken to cover the winter period in the United Kingdom.

Complete diet feeding

Instead of feeding forage (hay or silage) between milkings either on an ad lib basis at the silage face or in troughs in the 'loafing' area, and the concentrate or production part of the ration in the milking bail, many farmers with large herds are now mixing the whole ration and allowing the cows continuous access to the total of each day's feed.

Points in favour of this system:

1. Milking time is speeded up, because the occasional slow eater of concentrates is eliminated. It has been shown that there are inaccuracies in measuring the correct quantity given to each cow as she comes through the bail. The installation and upkeep of expensive dispensing equipment is avoided. Modern strains of dairy cows do not need the expectation of feeding in the bail to get them to come in for milking.
2. There is a saving of labour in not having to fork out silage and/or hay separately.
3. Much more accurate rationing of cows, especially if mixer wagons are used on which there are load cells to give accurate weighing, so that the amount for each day's consumption is known.
4. Improved management is encouraged by the necessity of having to divide the herd into at least three groups; the recently calved and heavy milkers, those on the downward slope of their lactation curves and dry cows. It is a convenient way to be sure that cows in the high-producing category get the high-energy and protected protein ration they require.
5. Surveys conducted by the Ministry of Agriculture show that oestrus detection has improved when farmers have changed to complete diet feeding. This is possibly related to the better body condition frequently noticed, itself a reflection of reduced weight loss after calving from better feeding.

The cost of equipment, such as mixer wagons, is high and a higher degree of managerial skill is demanded. While the results of surveys to date do not show a marked economic advantage from the system, increasing numbers of farmers with large herds are adopting it. Total yield may only be a little higher but many report higher milk quality ratings. This is likely to be of greater importance in future marketing patterns.

Flat-rare feeding

A marked departure from past accepted practice of trying to match intake to the individual output of each cow is the system of giving unrestricted access to high-quality forage, with concentrate supplement held at a predetermined level. This level is altered in relation to the stage of lactation in autumn-calving cows at infrequent intervals, perhaps with only one step-down in January.

This is a method that aims to make milk production more profitable by limiting the intake of energy from cereals and oil-seed meals, but maximizing consumption of high-quality cheaply produced or purchased bulky feedingstuffs. High quality silage produced on the farm is likely to be both the cheapest and the most generally available in the necessary quantity, but barn-dried hay, or brewers' grains can be used.

Silage should be of at least 25 per cent dry matter, cut and made with care, and there must be arrangements to allow a sufficiently extensive 'face' to be exposed to allow all milkers to eat whenever they wish, or blocks can be mechanically cut out and put into troughs of ample capacity. The silage should have an energy value of at least 10 MJ of ME per kg DM. If a concentrate (rolled barley 85 per cent, soya-bean meal 14 per cent, minerals 1 per cent) is given at between 6 and 9 kg DM per cow per day, the forage: concentrate ration can be kept to 65:35, and give a ration with an energy density of 10.85 MJ of ME per kg DM.

If the silage DM intake is 10 kg per day, the requirement for a 180-day winter feeding period will be 8 tones per cow of silage having a 25 per cent dry matter content.

It is claimed that if the composition of the concentrate given to the dry cows coming close to calving contains a proportion of maize meal and this continues to be fed to them after calving during the time of highest energy and protein demand, the degree of protected protein associated with maize and soya-bean meal will permit easier mobilization of her fat reserves, thus avoiding the danger of ketosis and assisting early peak yield.

Limitation of concentrate usage leads to lower peak yield but the steady intake of high quality forage ensures a flatter lactation curve.

Minerals

The milking cow's requirements are determined by her yield, since each kg of milk contains 1.19 g calcium and 0.99 g phosphorus. It has been shown that it is necessary to provide 2.2 g Ca and 1.5 g P/kg milk. If a requirement of 20 g Ca and 30 g P/day for maintenance is included, a total of 45 g Ca and 60 g P/day has been found adequate for cows with yields of 4500 kg in a lactation. Pasture normally contains Ca: P in a ratio 2:1 and cereals have an approximate Ca:P ratio of

1:4. Balance studies have shown that it is normal for a high-producing cow to be in negative calcium balance during the first part of her lactation. If extra minerals in the form of sterilized bone flour or calcium phosphate are added to the concentrates at a 1 per cent level, there will be no shortage of either major element. Under a normal feeding regime, shortages of sodium, magnesium or trace elements are unlikely. It is a customary precaution to include salt (NaCl) at 1.5 per cent of the concentrate ration. For a number of reasons, some cows out at grass in the early spring may have a low blood serum magnesium level (hypomagnesaemia). This is avoided by adding 50 g calcined magnesite per cow per day to the concentrates during this period.

Feeding Beef Cattle

Energy

The fattening of cattle occurs under a great variety of nutritional conditions, from the extensive grazing under range or pastoral management, where increase in weight depends upon seasonal rainfall, to intensive indoor feeding on high energy/low roughage feeding to maximize growth rate and attain marketable weight by about one year of age. In other words, the density of the energy in the feed, given by the metabolizable energy in each kilogram of dry matter (ME/DM) is a determining factor.

When any sort of forage (grass, hay, silage) is being digested and the products used for growth or fattening, more energy is lost in the process than when more concentrated feedingstuffs (cereal, oil seed meals) are being digested. In other words, the smaller heat losses of the latter means a higher net energy value. This is the basis of a useful way of calculating the amounts of different foods to provide a ration for cattle or sheep that are expected to make a stated daily gain in weight.

The animal production level (APL) is the ratio between the total net energy needed and that used for maintenance (APL = (Em + Ep)/Em, where Em = energy for maintenance, Ep = energy for grain). At the maintenance level, APL = 1; thus a table of APL values has been found for different liveweights and weight gains.

While the efficiency with which energy is used for maintenance can be taken as constant (at 0.72), it is otherwise with that used for weight gain. This depends upon the energy concentration (M/D) of the diet fed. An equation has been obtained for this variation expressed in net energy values and a table constructed.

Table 3.6. Animal production level (APL)

Liveweight, kg	*APL at liveweight gain (LWG) kg/day*					
	0.25	*0.50*	*0.75*	*1.00*	*1.25*	*1.50*
100	1.19	1.40	1.66	1.98	—	—
150	1.16	1.36	1.59	1.87	—	—
200	1.15	1.33	1.54	1.79	2.11	—
250	1.14	1.30	1.50	1.74	2.03	—
300	1.13	1.29	1.47	1.70	1.97	2.33
350	1.13	1.27	1.45	1.67	1.93	2.27
400	1.12	1.26	1.43	1.64	1.90	2.22
450	1.12	1.26	1.42	1.62	1.87	2.18
500	1.11	1.25	1.41	1.60	1.84	2.15
550	1.11	1.24	1.40	1.59	1.83	2.13
600	1.11	1.24	1.39	1.58	1.81	2.13

Table 3.7. Net energy allowances for maintenance and liveweight gain in growing and fattening cattle

Gain kg	*Net energy at liveweight, W, kg*										
	100	*150*	*200*	*250*	*300*	*350*	*400*	*450*	*500*	*550*	*600*
0	12.4	15.6	18.8	22.0	25.2	28.4	31.4	34.8	38.0	41.2	44.4
0.1	13.3	16.6	19.9	23.3	26.5	29.8	33.1	36.4	39.7	43.0	46.3
0.2	14.2	17.6	21.0	24.5	27.9	31.3	34.7	38.1	41.5	44.9	48.3
0.3	15.2	18.8	22.3	25.8	29.3	32.9	36.4	39.9	43.4	47.0	50.5
0.4	16.3	19.9	23.6	27.2	30.9	34.5	38.2	41.8	45.5	49.1	52.8
0.5	17.4	21.2	25.0	28.8	32.6	36.3	40.1	43.9	47.7	51.5	55.2
0.6	18.7	22.6	26.5	30.4	34.4	38.3	42.2	46.1	50.2	54.0	57.9
0.7	20.0	24.2	28.1	32.2	36.3	40.4	44.4	48.5	52.6	56.7	60.7
0.8	21.4	25.7	29.9	34.1	38.4	42.6	46.9	51.1	55.3	59.6	63.8
0.9	23.0	27.4	31.8	36.7	40.6	45.0	49.5	53.9	58.3	62.7	67.1
1.0	24.6	29.3	33.9	38.5	43.1	47.7	52.3	56.9	61.5	66.1	70.7
1.1		31.3	36.1	40.9	45.7	50.6	55.4	60.2	65.0	69.9	74.7
1.2			38.6	43.6	48.7	53.7	58.8	63.8	68.9	73.9	79.0
1.3				46.6	51.9	67.2	62.5	67.8	73.1	78.4	83.7
1.4					55.4	61.0	66.6	72.2	77.7	83.3	88.9
1.5						65.2	71.1	77.0	82.9	88.8	94.7

An example of how this system can be used, we may consider a beast, born in the autumn, reared as a calf inside during the winter months, then allowed to graze improved pasture during the spring and summer. As a yearling weighing say, 350 kg, it will be brought inside for the winter and required to gain weight at, say, 0.75 kg/day. If a supplement of rolled barley is to be fed at 2 kg/h/day and good quality grass hay is available, it may be important to know how much hay should be given each day.

From tables, it is determined that these two ingredients have:

	DM, g/kg	*ME/kg DM*
Rolled barley	850	13.0
Grass hay	800	9.0

The net energy requirement for a 350 kg animal with 0.75 kg gain per day = 41.6 MJ/day.

The APL factor = 1.45

The DMI should be approximately 2.5 per cent of liveweight, i.e. 8 kg

At APL 1.45	Barley	8.6 MJ/kg
	Hay	5.1 MJ/kg

	DM	Net energy, MJ
2 kg barley	1.7 at 8.6	14.6

Balance of net energy required: 41.6 – 14.6 = 27.0 MJ

Hay $\frac{27.0}{5.1} =$ 5.3 kg DM

	DM	Net energy, MJ
6.6 kg hay	5.3 at 5.1	27.0
	7.0	41.6

So 6.6 kg hay must be fed with 2 kg of rolled barley.

Protein

It is necessary to be sure that the feed mixtures given to fattening cattle contain sufficient protein. Except for cattle at the earliest stage of growth, it has been found that the protein content of a ration that will provide the energy for a desired rate of weight gain will generally be high enough to provide most, if not all, of the rumen degradable protein necessary and that this microbial protein will provide sufficient amino acids to sustain the level of growth desired. If there is a deficiency in the RDP for a particular level of ME intake, the addition of urea will provide the extra nitrogen found to be necessary.

In the example previously given, the protein required is set out as follows:

	Composition of ratio		
	RDP	*UDP*	*dg*
Barley (rolled)	86	22	0.80
Silage (grass)	136	34	0.80
	Requirements		
Maximum DMI 8 kg	546	4	
	DMI	*RPD*	*UDP*
Barley (rolled)	1.7	92	14
Silage (grass)	5.3	721	181
		813	195

This shows that there is in fact an excess of both RDP and UDP being provided. But this may not always be so if a low quality forage like straw is used.

Example

A steer of 300 kg live-weight is required to gain at 1.25 kg/day

ME requirement = 79.6 MJ/day

	Composition of ration				
	DM, g/kg	ME, MJ/kg	RDP, g/kg	UDP, g/kg	dg
Barley straw	860	7.3	30	8	0.8
Rolled barley	860	13.7	86	22	0.8
	Requirements				
	DMI, kg	ME, MJ	RDP, g	UDP, g	dg
	7.9	79.6	621	0	1.0
	Ration				
Barley straw	3.0	21.9	90	24	
Rolled barley	4.2	57.5	361	92	
	7.2	79.4	451	116	0.8

There is a deficit of 170 g RDP, which can be provided by 108 g/kg DM of urea (728 g/day)

Feeding Calves

Much evidence points to the desirability of calves suckling their dams as quickly as possible after they are born. The colostrum they obtain at this time is of importance in its content of immunoglobins, protective bodies against the bacterial contaminants of the environment into which the calf is born. In addition, colostrum is high in energy because of its fat content and helps the calf to clear the meconium from its intestine. The best husbandry practice is to extract up to three pints of colostrum from the newly calved cow (or from a stored supply) and feed it at once to the neonatal calf. Calves so protected have been shown to be much less likely to succumb to subsequent infection.

During the following four to six days, whole milk should be given whether they remain with their dams and suckle them or are taken away within 24-36 hours and are given the milk by teat or bucket. It should be fed three times daily at 41°C. Towards the end of this first week, the whole milk should be gradually changed to a milk replacer. This latter is a substitute milk bought in finely powered form and, when made up with the correct amount of water, has approximately the same nutritional value as whole milk from cows. It is in most marketing situations much cheaper.

The milk replacer

In European countries, this milk replacer will probably contain at least 60 per cent of dried skim milk powder, supplemented with additional fat, milk sugar, minerals, vitamins. It is important that the skim milk powder with which it is composed be prepared in a manner that does not damage the nutritionally important milk proteins. This is likely to occur during the manufacturing process if it is over-heated, through either a too-fast spray-drying process, demanding a very high level of heat, or during the older roller-drying process giving too long a time of exposure of heat. Low heat spray-dried milk powder may be a little more expensive, but is not likely to cause the indigestible abomasal clotting that occurs with cheaper milk replacers.

For some years there has been a tendency for calf-rearers to favour milk-replacers with a high fat content (15-20 per cent) for they provide more energy. Vegetable oils and animal fats (lard and tallow) are used in its manufacture, along with antioxidants.

Systems for Calf Feeding

The calves produced by the dairy industry are reared for three purposes: (1) to provide dairy cow replacements; (2) to be reared for beef; (3) to be marketed as veal. The latter represent the smallest proportion.

System (1) and (2): calves destined to be used for these purposes are commonly reared until weaning, and even to three months, in the same manner.

Milk replacer is expensive, so the aim is to make the period of its feeding as short as is commensurate with good growth in the calves. Many trials have shown that it takes a minimum of five weeks of milk feeding to have calves eating enough cereal-based concentrates, along with good quality hay, before they can be completely weaned.

Methods for feeding milk replacer

The traditional method of rearing calves is to feed each one from a bucket of warm milk twice each day. Modifications designed to save labour costs are to feed only once daily and to feed the milk cold or even, with calves over three weeks old, missing feeding milk altogether on one of the days of the weekend. These all tend to lead to slower growth and, unless the stockmanship is of the highest order, there is likely to be more digestive troubles.

Machines that mechanically mix set amounts of milk replacer powder at set times throughout the day from which calves suck from a single teat or series of teats have been available for many years. As labour becomes most costly, the capital expense of the machines and their installation in large units becomes economically justifiable. With such a system there is a much greater consumption of milk powder, with less concentrates eaten, making the cost of rearing calves to 5 weeks more expensive. Some comparative trials have shown that, as a result of the more frequent drinking of a uniform product, there is less digestive disturbance and a lower mortality rate in machine-fed calves.

Whatever method of milk replacer feeding is used, it is essential to see that the milk powder is thoroughly mixed with the water and in the correct proportion. With the commonly used 60 per cent skim milk powder replacer, it is stated that the best digestion of the casein clot that forms in the fourth stomach (abomasum) occurs at a concentration of 12 g per litre, which is equivalent to 12.5 per cent dry matter of cows milk, approximately the same concentration of total solids in the average sample of whole milk.

If the milk is to be fed warm, it is also very important that the temperature of the mixture when given to the calf is 40-42°C, which is higher than the calf's temperature. The calf likes it at this temperature, so drinks more and drinks quickly, in this way causing the maximum release of saliva, thus helping with digestion; while the rennet of the stomach, essential for clotting of the milk, is most active at 41°C.

By the end of the first week, the calf will be drinking 1-1.5 litres twice daily, and this should be increased by the end of two weeks to 2 litres at each feed. With individual feeding it is held constant at this amount, so encouraging the calf to eat increasing quantities of concentrates and hay. This concentrate mixture must be as palatable as possible; also of high nutritional value.

Table 3.8. Concentrate mixture for calves before and after weaning

Mixture constituents	*Percentage of constituent for calf:*	
	Before weaning, %	*After weaning, %*
Rolled barley	30	60
Crushed oats	15	20
Flaked maize	25	5
Molassine meal	10	—
Soya-bean meal	14	14
White fishmeal	5	5
Dicalcium phosphate	0.5	0.5
Salt	0.5	0.5
Vitamin A	25,000 IU daily	25,000 IU daily
Vitamin D	2500 IU daily	2500 IU daily

For convenience, many farmers prefer to use commercially available calf starter feeds, commonly made into small pellets or 'pencils'. The hay that should be put out for the calves in low racks from the first week must be of the highest quality. Only enough hay for the calves to clear up each day should be put out. Uneaten hay must be removed.

When the calves are eating at least 1 kg of concentrates daily for those of the light breeds and 1.5-2 kg for Friesian calves, the milk replacer mixture is given in decreasing amounts over 3-4 days, so they may be weaned as close as possible to 35 days of age.

Feeding after Weaning

As they approach the age of 10-12 weeks, the concentrate mixture is simplified and cheapened.

If the calves are spring-born, they will at this time be turned on to specially saved and thus relatively worm-free pastures. Autumn-born calves will have been turned out on to saved pastures as soon as weather conditions are suitable. The change-over from the hay-concentrate diet to grazing must be done gradually.

System (3): Veal calves

The calves may be kept under an intensive system, that is kept inside in single pens and fed wholly upon a high-fat milk replacer until 14-16 weeks old, being then sent for slaughter as 'white veal'. Less expensive veal but with a pink flesh is produced by allowing the calves access to concentrates, keeping them in small groups of similar size upon straw bedding. Feeding of high fat milk replacer continues until slaughter at the same age as 'white veal' calves.

FEEDING SHEEP

There are three periods in the breeding ewe's year when it is necessary to be sure to provide a supplement to what can at other times be simply maintenance level of feeding. These are: (1) just before and during mating (autumn); (2) during the last two months of pregnancy (late winter or early spring); (3) during lactation.

Most sheep are kept upon pasture throughout the year, but there has been an increasing tendency where winter weather is harsh, either to house the ewes throughout the worst winter months, or at least to bring them in for lambing.

As soon as this is completed, ewes and lambs are taken out to improved pastures. These will usually have received nitrogenous fertilizer in the early spring and will thus be able to provide ample highly nutritious grazing for both ewes and their lambs.

The techniques for more intensive forms of fat lamb production are understood, although rarely put into practice. This method demands that ewes be kept under cover and hand-fed for much of the year. Lambs are weaned 24-36 hours after birth, as soon as they have suckled their dams to obtain sufficient colostrum, and artificially reared on a milk replacer mixture and concentrates. Three crops of lambs in two years may be obtained by this system. Economic consideration make such a method impractical, wherever pasture can provide enough of this much cheaper feed for both ewes and lambs.

Energy

In the U.K., the same methods of determining the amount of energy necessary for the ewe's maintenance and subsequent pregnancy and lactation have been adopted as for dairy cows. This entails beginning with a knowledge of the basic expenditure of energy when the ewe is fasting, then agreeing the efficiency with which energy is utilized at the maintenance level. An activity allowance (0.15) is added, along with a safety margin of 0.05. This gives an equation for maintenance of 1.8 + 0.1 W (W = weight of sheep in kg). For a 60-kg ewe, this will be 1.8 + 0.1 × 60 = 7.8 MJ of ME per day.

Protein

The same consideration of the use of nitrogenous substances as discussed with dairy cattle will, in future, be given to deciding on the protein level for ewes. For the pregnant ewe, the requirement for undegradable protein is believed to be either absent or small enough to be ignored for practical purposes.

The requirement, as with energy, depends on the size of the ewe and, if it is known, whether carrying a single or twin foetuses. For a liveweight ewe (40 kg), the rumen degradable protein rises from 45 g per day at a maintenance level to 85 g per day at term; while for a 75-kg ewe receiving a diet of good digestibility, with a single lamb, the requirement is 75 g per day at maintenance and 130 g per day at term.

The mixture that is often fed to the pregnant ewe just before lambing, consists of good quality hay or high dry matter grass silage, supplemented with a cereal such as rolled or treated barley with 10-15 per cent of an oil-seed meal. This will provide ample crude protein to satisfy the ewe's requirements. It is only the large lowland ewe carrying twins that needs the higher level of oil-seed meal.

It is to be remembered that the capacity of the ewe carrying twins to consume forage is reduced during the last two weeks because of the pressure exerted by the uterus upon the rumen. It is necessary to make up the decreased energy intake from forage (some 20 per cent less) by giving more concentrates.

Minerals

While ewes are grazing ample pasture, it is unlikely that there will be any deficiency of a major mineral. This will not be so when sheep are kept inside or in yards and given cereals as well as roughage of varying quality. Since the cereals commonly used as stock feed (oats, barley, wheat, maize) contain approximately four times as much P as Ca, it is plain that with a desirable Ca:P ratio of 2:1, the feeding of cereals must be accompanied by supplementation with calcium. Calcium carbonate ($CaCO_3$, limestone) is a cheap source, but unpalatable unless combined with other substances. Commonly used are dicalcium phosphate or sterilized bone-flour. Common salt (NaCl) is used in making mineral mixtures, partly to provide additional sodium, but also to entire consumption. A common mixture that can be used with all ruminants eating diets in which cereals form a part is: $CaCO_3$, 25 per cent; sterilized bone-meal, 45 per cent; salt, 30 per cent. More costly but with increased effectiveness is a mixture of salt, 50 per cent; sterilized bone-meal, 25 per cent; calcium phosphate, 25 per cent.

In areas where there is a low copper content in the soil, especially when it is associated with higher than normal molybdenum and sulphate values, there is a likelihood of copper deficiency occurring in both ewes and their lambs. This is prevented by subcutaneous injection of 250 mg Cu in the form of an organic copper salt directly to the ewes or by adding copper sulphate fertilizer that is applied to the pastures.

In a number of countries there are areas where cobalt is deficient in the pasture, giving rise to a lack of vitamin B_{12} in sheep, a condition known as 'pine', 'coast disease', 'bush sickness' or 'enzootic marasmas'. Since the amount of cobalt needed is very small (1 mg per sheep per day), a convenient way of preventing deficiency is to give a small pellet of heavy cobaltic oxide by mouth. This lodges in the reticulum and releases its cobalt over several months.

Selenium is another element that is deficient in some soils and the pastures or crops that are grown on them. In New Zealand's South Island this has been shown to be the cause of 'illthrift' in sheep resulting in a low birth rate in lambs and an increased incidence of muscular dystrophy. If ewes are given 1-5 mg Se as either sodium selenate or sodium selenite once or twice, depending on the severity of the deficiency in the area, during the last month of pregnancy, and the lambs subsequently given supplementary dosing at intervals, prevention is achieved.

Mating

If the genetic potential of a ewe for reproduction is to be realized it has been repeatedly shown that she must be at her *optimum weight* for her breed or type when mated. Since ewes after weaning are often given poor pastures to graze, it is wise practice to turn them into improved pastures and even give supplementary cereals for a few weeks before joining the rams in the autumn. This means to the 60-kg ewe another 14 MJ of ME per day, in addition to her maintenance of 7.8 MJ of metabolizable energy per day.

The digestible dry matter of pasture in autumn falls quite sharply, even though, as a result of seasonal rainfall, grass appears green and lush. It is likely not to give more than 9.5 MJ per kg DM. If the ewe eats 0.5 kg pasture DM per day, thus providing 5 MJ, she will need to eat 1 kg of a simple cereal mixture (say, 85 per cent rolled or whole barley, plus 14 per cent soya-bean meal, plus 1 per cent mineral mixture) to make up a total of 21.8 MJ of ME per day.

Pregnancy

Towards the end of pregnancy, when the foetus is making its fastest growth, there is an increased demand for all food ingredients. This is clearly all the greater in a ewe carrying twins or triplets. If we take the needs of a small, medium and large ewe, each carrying a single (S) or twins (T), the energy requirements during the last two months of pregnancy are given:

Table 3.9. ME requirements per day of pregnant ewes kept outdoors

Liveweight		*Maintenance*	*ME required at week (before lambing)*				
			8	6	4	2	*last week*
	S		6.1	6.7	7.4	8.2	9.9
40	T	5.8	6.1	7.1	8.2	9.5	11.0
	S	7.8	8.0	8.8	9.8	10.8	11.9
60	T		8.1	9.4	10.9	12.7	14.7
	S	9.8	9.9	10.9	12.1	13.4	14.8
80	T		10.2	11.8	13.7	15.8	18.3

Lactation

The energy value for ewe's milk is taken to be 4.6 MJ per kg. Most of the milk is given during the first month of the normal 3-month lactation. For rationing purposes, ewes have been arbitrarily divided into the generally smaller and less productive hill breeds giving 100 kg and 120 kg milk for suckling single and twin lambs, respectively; while the heavier lowland breeds of sheep are assumed to produce 150 and 170 kg milk per single and twin lambs during the 3 months lactation.

From a calculation of daily milk yields, tables of the total ME allowance that should be provided for these two classes of sheep during each of the three months of lactation have been constructed.

A useful approximation is to remember that a ewe suckling a single lamb will require the same energy intake during the first month after lambing as she will during the week before she lambs and nearly half as much again if she is suckling twins; i.e. 20.3 MJ per day for a 60-kg ewe with a single lamb and 27.5 MJ per day for the ewe with twins. This level of feeding is approximately three times the maintenance requirement.

The result from trials in Australia and U.K. published since the appearance of the M.A.F.F Bulletin 33 (1975) have shown that the energy requirement of ewes during late pregnancy and early lactation are probably greater than the stated in Table 3.9.

It was Australian workers who discovered in the measurement of the peripheral blood concentration of 3-hydroxybutyrate, a product of the mobilization of body fat reserves, a reliable indicator of undernutrition in the ewe. Any level above 0.7 mmol per litre is a sign of this state of affairs, and this they found when feeding ewes according to the standards in Bulletin 33. Since then, Rowett Research Institute workers have confirmed that in ewes carrying twins and in such ewes during their first month of lactation, the degree of underfeeding could reduce the birthweight of lambs and cause a lower level of milk production. Robinson (1980) has suggested that maintenance levels of energy can more accurately be measured by use of the metabolic weight of the sheep, with the relationship:

$$\text{Energy for maintenance} \simeq 0.42\ W^{0.75}\text{kg MJ of ME in DM}$$

Recent studies have shown that the intake suggested for twin bearing ewes need to be increased by between 20 and 50 per cent at different stages of pregnancy. This appears to be the situation when the ration fed is at the common concentration of 10.0-10.5 MJ of ME per kg DM. If a diet of higher energy concentration is fed, the difference between the requirement stated in the M.A.F.F. Bulletin 33 (1975) and the Rowett figures is much smaller, due to the fact that the gross efficiency with which metabolizable energy is used by the conceptus appears to be a function of energy density.

Importance of Protein Quality in Pregnancy

A matter of the greatest importance that has emerged from the above work is an understanding of the vital role that protein, especially if it is not entirely degraded in the rumen (undegraded dietary protein, UDP), plays in making it possible for the ewe effectively to utilize her stored body reserves.

The ewe must call upon these if the feed intake during the last part of pregnancy is insufficient for the ewe carrying twins or triplets, and is even more important as an aid to high milk yield during the immediate post-natal period. At this time, it is very difficult for a ewe with vigorous twin lambs to have a large enough intake of metabolizable energy to meet lactational demands.

The practical implication is that every effort must be made to have ewes in optimum condition at the beginning of pregnancy by providing ample autumn grazing. As lambing draws near and, later, during lactation, it has to be realized that the ewe can only draw on her stored energy if there is sufficient absorption of amino acids from the intestine. In other words, it is not enough to provide the usual level of about 12 per cent crude protein in the dry matter of her diet if there is a likelihood of her energy intake being limited at any time.

When energy intake was reduced by 20 per cent, it was shown that nevertheless full lamb growth could result if the protein level in the ewe's diet was increased to 15 per cent. A comparison of various protein supplements and their cost, showed that fishmeal was the cheapest. Its rumen degradability is less than oil-seed meals.

Feeding Lambs and Hoggs

In some areas, the aim of the majority of fat lamb producers is to able to market a higher proportion of single lambs by the age of 16 weeks, leaving the lambs with their dams until this age. In upland flocks or where there is a proportion of twin lambs, weaning takes places between 10 and 12 weeks and the lambs are fattened either on hay aftermath, on improved pastures kept for the lambs after early cuts for silage, or on fodder crops. There is likely to be a varying need to provide supplementary feeding either with late lambs having to be finished on autumn pasture or with hoggs or hoggets being carried through the winter. These may be given access to root crops such as white turnips or swedes or to kale.

It may be decided to give limited access to such crops and to supplement their use with hay and a cereal concentrate, or this may be the only type of feed available. To calculate the amount required to attain a desired rate of gain, the same procedure as has been explained in Fattening of cattle should be undertaken.

If the Ministry of Agriculture's Bulletin 33 (1975) is consulted, it will be seen that the formulae have been translated into useful tables to show the net energy allowances necessary for growing lambs and the animal production levels for sheep as well as a table to show how the animal production level can be used to derive the net energy values, both for the maintenance and production it terms of ME per kg of DM. To use these tables and, therefore, to get a precise value for the amount of the different ingredients in a ration that should be fed, it is necessary to start with a knowledge of the weight of the lambs or hoggs that are being fattened and the food values for the ingredients that are available on any particular property. Hoggs of between 40 and 50 kg liveweight will not consume more than 1-1.25 kg of dry matter per day and their energy requirement will be satisfied by the consumption of about 0.75 kg of good quality hay with slightly less than this amount of a barley concentrate or compound feedingstuff which is likely to contain about 12 MJ per kg DM. The requirement for digestible crude protein for a hogg of this size gaining at 200 g per day would be 55 g per day. This is likely to be more than satisfied by the foods being supplied for energy purposes.

Use of Grassland

Dairy Cows

The most abundant and cheapest source of nutrients for ruminants in temperate climates is grassland. The extent to which it is utilized depends upon economic forces. If the costs of production of milk, those that are fixed, like buildings and interest charges on capital invested, and those that vary, like wages and the price of bought-in feedingstuffs, become high in relation to the price of milk, there follows a re-examination of grassland use. Where cows have to be housed for some 180 days during the winter months, with herds calving throughout the year, a traditional system of production is to try to maximize this by incorporating a large proportion of cereals and oil-seed meal in the diet.

When market forces turn against the producer, it is not difficult to show that where grass will grow reliably, in many countries this corresponds to the area where most cows are kept, the cost per litre of milk is reduced as pasture, both in fresh and conserved form, makes a larger contribution to the nutrient intake.

In the past, the higher winter price for milk has encouraged autumn calving, with the accepted convention that generous feeding of concentrates up to and beyond the lactation peak allowed cows to produce towards their genetic potential. There was the additional fact that, as the autumn-calving cows were put out to grass in the spring, a lift in the daily production could be anticipated.

In the U.K. it is shown that, when the cost of grazing is approximately one-fifth the cost of bought in concentrates (using as a basis the recent all-in costed price of 0.24p per MJ of ME for grazing and bought-in dairy nuts at £130 per tonne or 1.21p per MJ of ME), the dairy farmer anxiously examines ways to save concentrates without too severe a drop in milk yield.

The simplest way to do this is to have most of the cows calve in the spring, so that the heaviest nutritional demand as the cows come up to peak production can be met by high quality saved pasture. If it can be arranged that the herd obtain 85-90 per cent of its intake from pasture, supplemented with 400 kg concentrates for a full lactation (305 days), an average production of 4500 litres of milk has been found feasible.

The national average is not much above this; but when a majority of herds calve in the autumn and use only 50-60 per cent of total intake as grass products, there may be as much as two tonnes of concentrates being fed.

In the U.K. of late, there has been a steady and substantial move towards conserving pasture as silage rather than the traditional sun-cured hay. It is a move helped by the use of formic acid as a common additive to reduce fermentation loss during ensiling.

A trial in Northern Ireland with January-February calving cows that were given access to ample high-quality grassland, had an average production of 5500 l per cow. When the amount of concentrate was increased from 0.45 kg to 4.00 kg per cow per day, the mean daily milk yield only increased from 23 litres to 24 litres; but it meant that the total consumption of concentrates during the grazing period would have increased from 50 kg to 430 kg per cow. All that happens is that there is a pasture-sparing effect, the cows obtaining much of their energy from concentrates rather than from grazing.

Carefully managed and appropriately fertilized grassland, especially if cut three times in the season, can be made into high dry matter silage (25 per cent DM) and have a 'D' value (the digestibility of the organic matter, DOM, in the dry matter) of 65-66. It will then give 10.5 MJ of ME per kg DM and, if 7.4 tonnes of silage per cow of this high quality silage is available for the 180-day winter-feeding period, the exact level of concentrate supplementation to give maximum profitability must depend upon market forces that have current relevance to the circumstances of each dairy farmer. The important fact is that the cheapest source of energy and protein reside in the nation's grassland.

Sheep

A plan aimed at increasing fat-lamb production from grassland was produced by Newton and Young of the Grassland Research station a few years ago and given the name 'grasslamb'. It has now been successfully tried out on a small number of commercial fat-lamb producing farms over several years. It depends upon careful pasture management, with the use of nitrogen fertilizer at 251 kg per hectare (200 units per acre) and heavy stocking of medium-sized ewes producing almost two lambs per year.

The lambs are rotated through a series of small paddocks with forward creep grazing to give them the first fresh picking. It is important that the total area is large enough to provide sufficient conserved forage on which to over-winter the ewe flock.

It was found that even on farms where the level of management was already high, the 'grasslamb' system gave increases in output ranging from 4 per cent to 14 per cent. Carcass yields of 450 kg per hectare (equivalent to 1000 kg liveweight per hectare) were obtained.

Most of the lambs following stocking at 30-34 lambs per hectare can be marketed as fat, but there will be some that have to be sold as stores. The figures from a five-year trial of the system have been published and with carcass output ranging from 440 to 540 kg per hectare, the level of gross margin over the five years varied from £217 to £344 per head. It demonstrated what high stocking rate on suitable grassland can achieve.

Feeding Pigs

It is a well-attested fact that pigs are exceedingly adaptable to a variety of environments and feeding regimes. The fact is that next to the modern layer and broiler, the modern pig, with its simplified breed structure and widespread hybridization, has been intensively selected for rate of gain and efficiency of food conversion (often written FCE), the amount of food necessary to produce unit gain. In the 20-kg pig, this can be as low as 2 : 1, but gets progressively higher as the pig ages and lays down a higher proportion of fat. In the 100-kg pig, it has fallen to 4 : 1.

Growth rates after weaning can be as much as 1 kg per head per day, with a clear reflection of the composition of the diet and the amount fed in the composition of the daily weight gain. The greater quantity of energy consumed by a pig allowed ad lib feeding will simply result in the animal converting a higher proportion to fat than when the intake is controlled.

It appears that the maximum rate of lean growth for the most pigs is approximately 450 g daily. Any gain above that will be fat. The protein or lean that makes up the muscle mass of a pig is always associated with 80 per cent water, which is a reason why all rapidly growing animals need ready access to water all the time.

Even a young pig will be using energy to lay down fat in roughly equal proportions to the increase in lean. At the baconer stage, even though there is a restricted intake of energy, it will be laying down for times as much fat as protein. It has already been noted that at the same time the efficiency of food conversion is similarly diminished.

Energy

There are still many statements of the requirements for energy in the form of digestible energy (DE), which does not take account of the energy loss in the urine. With the accumulation of data on the digestibility of the various feedingstuffs given to pigs, there has been a move towards the more precise metabolizable energy (ME1/4) unit.

For the cereal-based diets commonly fed, where the cereal protein is likely to be supplemented by oil-seed meal or fishmeal, the ME value is approximately 95 per cent of the DE value.

In the same manner as has been considered in the feeding of other farm animals, the calculation of the amount to be fed is most accurately done by first considering the requirement for maintenance, then adding to this figure the production requirement, the extra energy, protein, minerals that the particular group of pigs need for growth, pregnancy or lactation. It is widely accepted that 80 per cent of the cost of producing pig meat derives from the food, thus emphasizing the wisdom of care in rationing.

Maintenance requirements

This is most accurately calculated by relating it to the metabolic rather than the actual weight of the animal. The basic metabolic rate upon which maintenance is based is a function of the surface area and this is expressed as the metabolic weight, $W^{0.75}$.

Sows

During the early part of pregnancy the maintenance need is believed to be reduced to permit body reserves to be accumulated. Most sows need to do this to a greater or lesser extent according to their age, and thus the number of preceding pregnancies and the feeding regime previously followed have to be considered. During the later part of pregnancy there is loss of heat from the increased metabolism of the uterus and the utilization of the sow's fat reserves. The following have been suggested.

Early pregnancy: 0.40 MJ of metabolizable energy (ME) per kg $W^{0.75}$

Late pregnancy: 0.55 MJ of ME per kg $W^{0.75}$

A single figure of 0.475 MJ ME per kg $W^{0.75}$ or 0.5 MJ DE per kg $W^{0.75}$ for the whole of pregnancy can be used.

For a sow weighing 150 kg, this would be 20.3 MJ ME, or 21.4 MJ DE per day.

Growing pigs

Maintenance for the growing pig is also taken as 0.475 MJ of ME per kg $W^{0.75}$ or 0.5 MJ of DE per kg $W^{0.75}$.

It has been found that growth of lean demands 15 MJ of DE per kg of lean formed, while fat requires 50 MJ of DE per kg formed. If the proportion of lean to fat is calculated according to the weight of the grower, maintenance will demand, 5.3, 8.9 and 15.0 MJ of ME at 25, 50 and 100 kg, respectively.

Production requirements

Sows

Allowance must be made for the need of the sow to gain weight during pregnancy, the energy cost of which is taken to be 25 MJ of ME per kg gain. If the total gain is 18 kg over a 115-day pregnancy, the extra energy is (24 x 18)/115 = 3.91 MJ daily. For the sow of 150 kg mentioned above, this would mean a daily requirement of 24.2 MJ of ME. The type of cereal-based mixture that is now fed yields 12.5 ME per kg. Thus sows of this size would be fed close to 2 kg per day.

Lactating sows must be fed according to their weight and the number of piglets they are suckling. The milk of sows has twice the amount of protein and fat as cow's milk. Its energy value is 5.4 kJ per kg. The efficiency of conversion of dietary energy to milk energy is 65 per cent, making it necessary to provide 8.3 MJ of ME for every kg of milk produced.

It is normal for the sow to call on her fat reserves to provide milk, especially in the first three weeks as peak production is reached. In fact, it is impossible for a sow suckling 8 or more piglets to consume and absorb sufficient energy. It has been estimated that 1 kg of weight loss will yield 28-29 MJ of milk energy. In doing so, it will save 44 MJ from the diet.

It must be emphasized again that it is important for the sow to get enough food during pregnancy to permit her to begin suckling with the help of the fat reserves she has built up for this purpose. She must then be fed adequately during lactation according to the number of piglets being suckled. This will permit the sow to begin her next pregnancy (approximately 5 days after weaning) at optimum weight.

For example, a 150-kg sow with 9 piglets giving 7 kg milk daily would need:

Maintenance, 42.8 × 4.75	=	20.3 MJ of ME
Production, 7 × 8.13	=	58.1 MJ of ME
		78.4 MJ of ME daily

A commonly fed cereal mixture is likely to contain 12.5 MJ ME per kg (or 13 MJ DE per kg). To prevent weight loss in this sow, it will be necessary to feed 78.4/12.5 = 6.27 kg per day.

An approximate guide suggested by one authority is to feed a ration of 4 kg per day of a 13 MJ of DE mixture for the sow and 5 piglets, and 0.4 kg more for every additional piglet. This must be modified according to the size of the sow and the efficiency with which it converts food.

There is also the fact that during the first few days after farrowing, the sows appetite is diminished. It is wise at this time to feed smaller quantities at least three times during the day.

Feeding after weaning

If early weaning at 2-3 weeks occurs, sows should still be in good condition so long as the feeding during pregnancy has been adequate. Weaning at 5-6 weeks may find some older sows too thin and they will have to be given generous feeding before and over the mating period.

Later weaning at 8 weeks should find the sows at their correct weight. A daily ration of 2.5-3.0 kg of standard mixture will be sufficient.

Protein

Unlike the ruminant, with a range of essential amino acids becoming available from the digestion of a considerable amount of microbial protein, there is, with the pig, a problem of getting into the ration the right balance of the essential amino acids.

The proteins of the cereals that make up 80-90 per cent of the diet fed to most pigs are deficient in lysine, and could be limiting in methionine + cystine and threonine. In recent years, soya bean meal has become widely used and can help to rectify the situation because it contains nearly twice as much lysine as barley protein.

The available pure synthetic amino acids are often more expensive than the source of a range of natural amino acids, such as can be found in soya-bean meal and fishmeal; hence in mixtures fed in most European countries, these ingredients are often included.

Protein requirements

Sows

Gilts are mated long before they have attained their mature size. It may not be until their fourth litter that this limit of growth is reached.

Protein is used in the foetuses and membranes and in building tissue reserves within the body and mammary glands. The amount of protein deposited in the uterus rises from 3 g daily at 15 days to 50 g daily during the last days before birth. Including the amount being laid down in the mammary glands, the total protein utilized during the last month is 65 g. To this must be added the amount being laid down in the sows body, in preparation for lactation, some 26 g daily.

There is an increase in the efficiency with which the sow uses protein during pregnancy. It is therefore concluded that 170 g digestible crude protein (DCP) daily is sufficient. This can be met if the level of protein in the ration is at the level of 14 percent digestible crude protein with the ration being fed at a rate to provide enough energy.

Lactation

The need for protein, as for energy, depends upon the number of piglets being suckled. There is 6 per cent protein in sow's milk and, with an average litter, the output of milk 5-7 kg daily. If it is 7 kg and there is 70 per cent efficiency of conversion from food protein, the sow will need 600 g of digested protein + 86 g digested protein for maintenance + 70 g digested protein to give slight body gain, a total of 750 g DCP.

Growers

Maintenance is taken to be 0.09 per cent of the weight, i.e. in a 50-kg pig = 45 g protein.

Lean muscle growth contain 22 per cent protein (the remainder is water), and if 450 g lean is put on daily (22 + 450)/100 = 99 g, making the total requirement 99 + 45 = 144 g body protein.

The biological value, i.e. the amount actually absorbed, of most dietary proteins is 65, so the digested protein = 144/0.65 = 222 g.

Most dietary proteins have a digestibility coefficient of 80 per cent; so the digestible crude protein (DCP) = 222/0.8 = 277 g.

As with the sow, it is found that in the practical day-to-day feeding, if the level of DCP in the mixture fed is kept at the level of 14 per cent and is made up of sources able to provide the correct levels of amino acids, the daily intake will be safeguarded so long as the energy intake is controlled.

Minerals

Calcium and phosphorus

The two major minerals that have to be provided in sufficient amount and in close to equal proportions in the diet are calcium and phosphorus. This is because they are the elements that are the main components of bone. Remembering the large amount of cereal eaten, there is the problem of making good the deficiency of calcium, since cereals contain up to four times as much phosphorus as calcium (0.33 per cent P and 0.04 per cent Ca). Much of the P is in the form of phytate and has reduced availability. Nevertheless the greater need is to provide extra calcium.

In starter and creep mixes, it is customary to add calcium carbonate and dicalcium phosphate so that there is close to 0.9 per cent of each in the complete diet. With the usual ingredients, this is likely to mean adding 12 kg per tonne of calcium carbonate and between 1 and 4 kg of dicalcium phosphate.

It is concluded that the growing pig needs 16 g Ca and 12 g P daily, and in the preparation of a complete diet, this will be provided at 0.8 per cent Ca and 0.6 per cent P, obtained by adding calcium carbonate and dicalcium phosphate at approximately 17 kg and 11 kg per tonne of feed.

Salt

There is need to provide additional sodium but not potassium. Sodium chloride (common salt) is commonly added at a level of 0.3 per cent. It has the additional advantage of acting as an aid to appetite. However, pigs are susceptible to salt poisoning, likely to occur if the dietary level is increased to 1 per cent.

Iron

The newly born pig has an inadequate store of iron in its liver because of the size of litters and shortness of the sow's gestation. With pigs kept inside, it is necessary to give additional iron to make up their requirements of 7 g per day. The commonest way now to do this is to inject an organic iron preparation intramuscularly within the first week of birth.

Other minerals

No other mineral deficiencies, either of major or trace elements are likely if the above supplements are carefully added to the common mixtures of ground cereals and protein supplements.

Copper has in the past 20 years, however, frequently been added to diets of fattening pigs at as high a level as 250 mg per kg of food to act as a growth stimulant. Since there can be high residual levels of copper in the livers of such pigs, the accepted level is now 175 mg per kg of food. Even at this level, it is important to be sure that sheep gain no access to such copper-supplemented concentrates because of the susceptibility of this species to copper poisoning.

Vitamins

In the rapidly growing young pig, there is a need for sufficient fat-soluble vitamins to be in the diet. This is especially true for retinol (vitamin A) and cholecalciferol (vitamin D). The stated requirements are :

Vitamin A. ... 1400 IU per kg food for growing pigs
3600 IU per kg food for pregnant and lactating sows

Vitamin D .... 230 IU per kg food for growing pigs
160 IU per kg food for sows.

Vitamin-E (α-tocopherol) is likely only to be wanted in a very small quantity unless any oil included in a mixture contains a high proportion of unsaturated fatty acids, or there is rancidity in the vegetable or animal protein supplements. Long-stored grain or grains that has over-heated during grinding and pelleting may cause a deficiency of this vitamin. Its antioxidant properties can thus be valuable, so that is often included in the pre-mixes of vitamins and trace elements that have become the common method of insuring against any deficiency or imbalance in the ration.

This pre-mix is commonly added at 0.3 per cent (3 kg per tonne) to give a vitamin E level of 10,000 IU per tonne (10 IU per kg). In the pig, as in ruminants, there is a close relationship between the action of selenium and vitamin E in the prevention of muscular dystrophy (white muscle disease), thus emphasizing the wisdom of including this vitamin at a high enough level.

Water-soluble vitamins

The pig has the capacity to synthesize most of the water-soluble vitamins in sufficient amounts if they are not contained in the mixture of feedingstuffs given. However, with long storage of some grains or when there has been overheating or careless processing, there may be a deficiency of *biotin*, revealing itself as softening and ulceration of the hooves and weakness in the hind limbs. There has been a tendency for the incidence of this deficiency to increase. The level of supplementation to be sure of avoiding signs of deficiency is 10 μg per kg and should form part of the pre-mix.

Practical Feeding

In the main pig-producing parts of the United States and other temperate countries, stock-feed producing companies make use of linear-programming techniques with their computers, continually to monitor the costs and nutritional content of the ingredients available to them in order to make least-cost diets. This has the advantage of making available to producers pig feeds competitively priced but whose composition varies. If this is of too great a magnitude, it can affect intake and cause a degree of nutritional stress that is undesirable.

Many producers, especially if they can grow their own cereals or can procure them locally on contract, prefer to mill and mix their own pig feed, adding a pre-mix of minerals and vitamins.

Apart from copper (added in the form of hydrated copper sulphate, $CuSO_4.7H_2O$), some pig producers believe that they get a growth response from the addition of an antibiotic. In many countries the antibiotics that can be added to feedingstuffs without a prescription from a veterinarian are severely restricted. In the U.K., only bacitracin, virginiamycin and flavomycin are in this category. It is also permitted to add compounds of arsonic acid and other arsenicals and these are sometimes added to diets for growers at up to 100 g per tonne. To prevent tissue residues, they must be withdrawn 10 days before slaughter. Where since dysentery exists, such antibacterial agents as nitrofurazone, furazolidone or tylosin may be added to feed when prescribed by a veterinarian.

Creep Feeds

In order to assist growth at as young an age as possible, when the maximum proportion of lean is laid down in the young pig, and food conversion is at its most efficient, it is the common aim to get the piglet on to some concentrates by the end of its first week.

To entice the piglet to eat this supplement, it has to contain ingredients that are both very palatable and of high nutritional value. Such are skim milk powder, sugar or molasses and flaked maize.

A satisfactory mixture to put into the creep for the first month would be the following:

	kg. per tone
Ground barley	70
Ground wheat	250
Flaked maize	200
Wheat feed	64
Skim milk powder	100
White fishmeal	50
Meat-and-bone meal	30
Soya-bean meal	170
Molasses	50
Calcium carbonate	13
Dicalcium phosphate	1
Vitamins and trace elements	3

This mixture has an energy value of 143 MJ DE per kg DCP of 200 per cent

The time during which such a relatively costly ration is fed need to be kept to a minimum. By the time the young pig is about one month old, a reduction can be made in the more expensive ingredients.

Such a modified creep feed for the young pig that will get it over the weaning stage is as follows:

	kg per tonne
Ground barley	183
Ground wheat	250
Ground oats	89
Flaked maize	100
Wheat feed	100
Soya-bean meal	93
Decorticated groundnut meal	30
White fishmeal	63
Meat- and -bone meal	30
Calcium carbonate	2
Dicalcium phosphate	4
Vitamin and minerals	3

Energy value – 140 MJ DE per kg DCP of 16 per cent

Grower and Finishing Rations

There has been an increasing tendency to simplify breed structure in any advanced pig producing countries, but it is necessary to be familiar with the growth and fattening characteristics of the particular pigs in the environment provided on each form before the correct energy density and daily quantity for each pig can be accurately determined.

For economic as well as nutritional reasons, there has been increasing use of high energy density diets, that is mixtures having digestible energy values of 13 MJ per kg or above. It means that there will have to be a proportion of wheat and/or maize, as well as barley included because of the higher energy value they contain (barley 12.7, wheat 14.0, maize 14.5 MJ DE per kg). Before deciding upon the quantities of each cereal for a grower or finishing mixture, the unit cost of energy should be calculated, that is how much does each MJ of DE, cost for each cereal. It is of course incorrect to attribute all of the cost to each nutrient, but, for purposes of comparison, this is a useful device.

A mixture for growing pigs, taken to be between 40 and 70 kg weight, that is made up of 70 per cent barley, 8.5 per cent wheat, 13 per cent soya-bean meal and 5 per cent groundnut meal will give 13 MJ DE per kg and 14 per cent DCP.

Rationing

Various rationing methods are pursued. The early period of ad lib feeding may extend up to porker weight if (60 kg) they are being marketed at this stage, but with baconers, limitation to intake to restrict the laying down of excess fat generally starts at 30 kg. The scale may be based upon periodic weighting or upon a time scale. One based upon the latter is to feed at 2 kg daily for the first month, increase to 2.5 kg for the next month, then 3 kg until finished at approximately 100 kg. So long as there is no deficiency or imbalance in the growers' diet, the energy: protein ratio that can be kept in mind to obtain a satisfactory rate of growth along with an acceptable fat cover, is one of 24 MJ: 280 g DCP daily. A high energy diet will supply this in as little as 1. 75kg (mixture containing 13.7 MJ DE per kg and 16 per cent DCP), while it would be necessary to feed as much as 2.25 kg of a low energy: low protein mixture (10.6 MJ DE per kg and 12.4 per cent DCP).

Pregnant Sow Ration

The wisdom of having an arrangement to allow sows to be fed on their own is widely recognized. Even when this is being done, there is

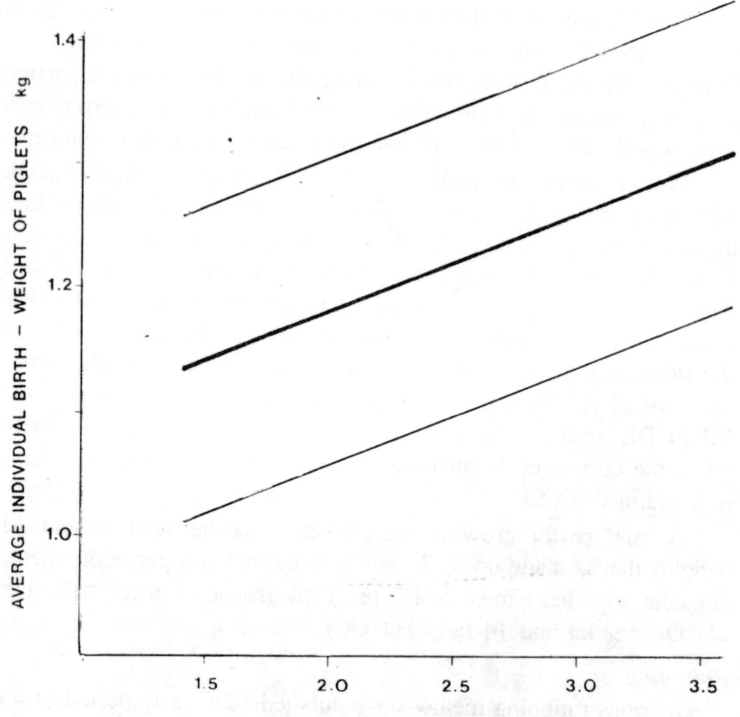

Fig. 3.4. The effect of level of feeding during pregnancy upon birthweight of piglets.

much individual variation in the efficiency of food utilization, added to which there is the effect upon body weight from suckling litters of varying sizes and the length of time during which the sow has suckled her litter. It is when the young pigs are weaned at 3-4 weeks that the sow is left in poorest condition, for she has had no time to make up the tissue loss inevitably associated with the steady increase of milk production to its peak towards the end of the third week.

The level of feeding before mating, then, has to be determined by these considerations. Assuming that a weighbridge is available, a convenient time to weight sows is when they are out of their pens for mating. It should be the aim to have sows gain in weight to the extent of between 12 and 15 kg during the cycle of one mating to the one following.

Both the energy density and protein level can safely be lowered during pregnancy. This will be achieved with a mixture of :

	per cent	
Ground barley	20	
Ground wheat	20	
Wheat feed	45	12.5 MJ DE per kg
Soya-bean meal	13	10.0 DCP per cent
Calcium carbonate	1.4	
Dicalcium phosphate	0.3	
Minerals and vitamins	0.3	

If a minimum ration of 2 kg of this or similar diet is fed during pregnancy, there should be a sufficiency of necessary nutrients.

It has also been shown that increasing the digestible protein intake beyond about 220 g daily has no increased beneficial effect upon the litter.

Lactating Sows

It is in lactation that there is a much more direct relationship between feed intake and the amount of milk produced, and also upon the extent of the sow's live weight loss. It is therefore necessary to feed according to the number of piglets that have to be suckled. A sow with a litter of 10, assuming she is approximately 140 kg in weight and losing 10 kg of body tissues daily, will have to be fed just over 6 kg daily of the same mixture that she was given during her pregnancy, for the requirement is 77 MJ DE daily. Some advocate a 1 MJ DE mixture with 12.5 DCP per cent. Less has to be fed each day.

Feeding Boars

Young boars have by no means finished growing but their intake must be limited to prevent excessive fat deposition. If they are given the same diet as the lactating sow, requirements will be met by feeding 2.75 kg daily. This must be adjusted according to the extent of his use which will be reflected in his weight.

Feeding Poultry

Fowl

There are two sharply different types of production: (1) laying hens; (2) broilers. The former must be raised from the day-old chick to the commencement of egg-laying at 18-20 weeks old, during this time being known as a 'replacement pullet'. The laying hen may be kept for either one or two years. A light hybrid laying hen is expected to lay 240-280 eggs in its first year. If kept for second year there is a custom of submitting the birds to a forced moult involving severe restriction of feed for 3-5 days.

The meat type of broiler bird, under optimal conditions of feeding and housing, will reach a marketable weight of 1.75 kg in 7-8 weeks.

Factors affecting appetite

While severe imbalance or shortage of an essential amino acid, vitamin or mineral will have effects on a bird's health and hence upon food intake, there are two factors that affect appetite. The most important is the concentration of energy, so that an increase will cause a decrease in appetite and vice versa. This has given rise to the saying that a bird eats for energy.

The other main factor is the environments temperature. As it rises the bird eats less for there is less loss of energy as heat. Over 30°C the appetite falls faster as greater difficulty is felt in getting rid of heat.

Other things that affect intake are the manner of presentation of the food, whether as mash, pellets or 'crumbles', for birds will more readily eat the latter two forms than dry, fine mixtures. Of the many trials made, it appears that a difference of about 7 per cent in favour of pellets represents a mean value.

The number of hours of light to which bird are submitted has an effect upon intake, especially of younger birds. This is the reason why broilers are normally given 23 or 24 hours light each day.

Finally, one should remember that viral, bacterial or parasitic disease can seriously affect appetite, even at a subclinical level.

Broilers

Since maximum growth rate is demanded, the concentration of energy in the diet must be high. Normally two diets are fed, the first or 'starter' mixture is fed from day-old to 4-5 weeks and a 'finisher' ration until marketing.

A survey in the U.K. in 1970 showed that the leading producers were feeding diets at 12.3 MJ of metabolizable energy (ME) per kg and finishing rations at 12.7 MJ of ME per kg.

A more recent study amongst broiler producers in the U.S.A. showed that while most used a starter ration with energy concentration at 13.6 MJ of ME per kg and a finishing ration of 13.8 MJ of ME per kg, experimental data were used to show that with starter : finisher diets ranging up to 14.4 : 14.7 MJ per kg, the weight at 54 days could be increased to 1776 g from the usual 1725 g at 54 days. The feed eaten was reduced from 3480 g per bird to 3300 g per bird as a result of the food conversion efficiency (kg gain/kg food) improving from 1.97 to 1.86. It is customary to include a quantity of vegetable oil (e.g. maize oil) and/or animal fat (e.g. tallow) in broiler diets to increase energy density. This varies between 15 to 20 g per kg as a minimum and may reach 50 g per kg.

Protein

It must be said only when there is a sufficiently high level of protein with the correct balance of essential amino acids, and of minerals and vitamins, can such concentrations of energy achieve these results.

Much effort has been put into discovering the correct levels of those amino acids which in the feedingstuffs available in a particular country, are likely first to be 'limiting, i.e deficient. In diets in the U.K., where ground wheat and ground barley, as well as maize meal, are used, lysine is the first limiting amino acid. It has been established that from 0-4 weeks 0.85 g per MJ is necessary, and in a finishing ration it should be 0.61 g per MJ. It is to guard against the likelihood of this amino acid becoming deficient that fishmeal, a rich natural source of lysine (having some 50 g lysine per kg), is almost always included in the rations for growing birds.

Expressed in another way, it is necessary for broiler starter diets to have 23 per cent crude protein of which lysine is 1.20 per cent when the energy value is 13.06 MJ per kg.

The protein level can be reduced to 20 per cent between 3-6 weeks and to 18 per cent, with lysine at 0.85 per cent, in finishing diets.

Minerals

Of the major elements, it is important that the calcium level does not exceed 10 g per kg, and is better at 8 g per kg; phosphorus should be 5 g per kg. Salt is normally added at a low level of 2-5 g per kg and must not exceed 8 g per kg. Manganese is an important trace element for rapidly growing chicks, for a deficiency can cause 'slipped tendons', i.e. the tendon of the gastrocnemium muscle slips off the trochlea of the tibia because of incomplete bone modelling. The absorption of manganese from the intestine is low at only 5 per cent, and an excess of calcium can further reduce its absorption. The requirement is 50 mg per kg for chicks.

The only other trace element which has been shown to be associated with a clinical deficiency state is selenium. A severe deficiency will cause exudative diathesis in chicks and its presence in adequate amounts can act synergistically with vitamin E or other antioxidants to prevent encephalomalacia. Recently it has been shown that shortage of selenium reduced the enzyme glutathione peroxidase and that such low levels make a bird more susceptible to coccidiosis. The supplementation should only be at the low figure of 0.1 mg per kg.

Vitamins

It is important that young birds be given a supplement of vitamins A and D of the fat-soluble group, since the ingredients of diets fed to young birds, ground cereals, fats and proteins from oil-seed meals, fish and meat meals, contain low levels of both the provitamins and preformed vitamins.

Retinol (vitamin A) should be added at rate of 1500 IU per kg of diet, but many commercial pre-mixes may have values several times greater than this. The incidence of avitaminosis A is very low nowadays, but if there has been an inadequate intake of retinol, the signs are to be found in the nasopharynx. The stratified epithelium that replaces secretory epithelium will show as raised white areas. There are often ocular lesions, conjunctivitis and keratitis; urates become deposited in the ureters.

Vitamin D_3 (cholecalciferol) must be added at the rate of 400 IU per kg feed, when the amounts and ratio of Ca : P are optimal. A deficiency of the vitamin leads to interruption of bone calcification or rickets.

If there is a low level of unsaturated fatty acids in the ingredients of a growing bird's diet and an antioxidant, such as ascorbic acid, ethoxyquin or diphenyl- p-phenylene diamine is included, the requirement of a-tocopherol (vitamin E) is minimal and no more than 10 IU or 50 g dried yeast per kg need be added.

Of the water-soluble vitamins, it is important that riboflavin is added, for there is evidence that broiler chicks have a higher demand than do the lighter white hybrid chicks that will grow into layers. Synthetically produced riboflavin should be added at 4.0 mg per kg. Deficiency of this vitamin shows itself, not only in the reduced growth and unthrifty appearance common to many deficiency states, but in the specific abnormality of 'curled toe' paralysis, an inward drawing of the claw, progressing to paralysis because of degeneration of the sciatic nerves.

With the sometimes lengthy storage and greater degree of processing of cereals that occurs now during the preparation of pelleted poultry feeds, it is becoming clear that vitamins, like pantothenic acid and biotin, can be reduced to too low a level. With a deficiency of pantothenic acid there is a sticky exudate from the eyes so that they remain closed; later there is a scaley dermatitis of the feet. The requirement is for 10 mg per kg diet.

Particularly when the diet of growing birds contains wheat and barley, rather than wholly being made up of maize and soya bean (as in the U.S.A.), is there a possibility of a deficiency of biotin. Much of the biotin is wheat and barley, and in fishmeal, is in non-available forms. The requirement is 0.25 mg per kg. Deficiency is associated with the fatty liver and kidney syndrome and may also cause a dermatitis under the feet with cracking and bleeding.

Additives

Many of the commercially available vitamin and mineral pre-mixed have one of the antibiotics added that are permitted, i.e. bacitracin, virginiamycin or lincomycin. Addition of one of these at 120 g per tonne is thought to aid growth.

A coccidiostat may be included. These change as the effectiveness of a drug diminishes. Zoalene and amprolium are two that have proved effective. Monensin sodium is much used at the present time, included at 115-120 g per tonne. Antioxidants are commonly included to protect fats and oils and fat-soluble vitamins. The include butylated hydroxyanisole, butylated hydroxytoluene and ethoxyquin. The latter at 0.125 per cent will improve the stability of xanthophylls used for yolk colouring; butylated hydroxytoluene is usually included at 0.02 per cent.

Pullets and Broiler Breeders

The type of diet used in the early stages of the growth of replacement pullets or stock being raised for the production of eggs for the broiler industry is similar to that given to broiler chicks, except that the energy and protein levels are a little less.

From the age of about 8 weeks, it is customary for the energy/protein levels to be further reduced for this is the beginning of a period of less rapid growth. Many producers rearing pullets and all who rear broiler breeding hens use diets containing 12. 2 MJ of ME per kg and 15 per cent crude protein. From 14 weeks to 20 weeks, taking the latter age as point of lay, the protein level may be further reduced to 12 per cent.

In areas where brown eggs are wanted, making necessary the rearing of somewhat heavier types of hybrid bird, likely to weight approximately 2.0 kg at 18 weeks, rather than the 1.6 kg of a White Leghorn type hybrid, there may be merit in limiting the intake of energy to 70 per cent of ad lib intake. The effect is not only to save on feed cost but by delaying slightly the onset of egg laying to reduce the number of small and over-large eggs that most birds lay during the first few weeks. Many of the trials that have studied this type of feeding regime have shown that while mortality during the period of restricted feeding may be slightly increased, there is a reduction in mortality during the subsequent laying period.

Layers

Most types of broiler breeders have a propensity to become too fat during the growing period, so there is a more general acceptance of the system of limiting energy intake with this class of bird.

During the laying period, the commonly accepted feeding practice is to permit feeding to appetite on a twice or thrice or daily operation of mechanical provision of a fine dry mash. Since most layers are now kept in battery laying cages, it is believed that the longer time taken by the hen to eat dry mash, rather than pellets, reduces the likelihood of the vice of feather pecking leading to cannibalism developing.

It has been found that the various strains of layers now being used will produce at an optimal level at an energy concentration of 12 MJ of ME per kg, with crude protein at 15 per cent. Slightly higher or lower energy concentrations are likely to give equally good results, so long as the protein is similarly altered and contains the correct amino acid balance. If the concentration of metabolizable energy is brought below 9.6 MJ per kg, egg laying is reduced.

Protein

Taking the average weight of an adult laying hen at 1.8 kg and the egg output at 50 g per day, it has been shown that the requirement for lysine is 7.5 and of methionine + cystine 4.7 per kg. This is on the basis of an intake of 110 g of mash per day and gives a total protein consumption of 18 g crude protein.

Minerals

The requirement for calcium for maximum egg production is 30 g per hen per day. The most satisfactory results for numbers of eggs and good shell thickness seems to be achieved when calcium is included at 3.7 per cent of the ration and phosphorus at 0.6 per cent, and with an adequate intake of vitamin D_3 (cholecalciferol) of 100 IU per kg diet. The commonest method of supplying the comparatively high levels of calcium needed is by adding to the mash the required amount of finely ground calcium carbonate. This appears to be relatively quickly absorbed, for it has been shown that most feeding occurs between 9.0 a.m. and 7.0 p.m. before the birds perch for the night. The shell of the egg to be laid next day is put around the egg during the early hours of the morning when the bird's blood serum calcium is at a low level. If calcium is provided partly as oyster shell grit this is more slowly absorbed and shell strength can be maintained more certainly.

Other minerals should be at the levels discussed for the ration of the growing bird. It is necessary to be sure that laying hens have a sufficient intake of retinol (vitamin A). It should be added to the dry mash diet in a stabilized form at the level of 2700 IU retinol per kg diet.

Of the water-soluble vitamins that are likely to be deficient in the diet of layers, riboflavin is the first that is likely to be limiting, especially if it is necessary to have fertile eggs. A deficiency of riboflavin leads to death of the developing embryo at about the 11th day of incubation in a high proportion of fertile eggs. While a level of 2.5 mg per kg is sufficient for commercial (unmated) laying flocks, breeders hens need 4 mg per kg of this vitamin.

In the mixes of cereals and oil-seed meals that are commonly used, there should be no deficiency of other water-soluble vitamins. If, through the use of long-stored grain and overheating during processing, there are any signs of pantothenic acid or biotin deficiencies, these vitamins should be included at levels to supply 1.5 mg per kg to layers and 6.5 mg per kg to breeders for pantothenic acid and 0.1 mg per kg and 0.15 mg per kg of biotin for layers and breeders, respectively.

An illustration of a practical layers mix which assures the non-availability of grass meal as a source of carotenoids for yolk-colouring.

	Composition, %	
Maize meal	43.4	
Wheat meal	20.0	
Middlings (weatings)	15.0	Energy: 11.64 MJ/kg
White fishmeal	1.2	Protein: 15.2%
Soya-bean meal	8.0	Lysine: 0.7%
Meat an bone meal	4.4	Methionine: 0.32%
Limestone flour	7.6	Ca: 3.5%
Methionine	0.058	Available P: 0.4%
Pigment	0.04	
Mineral/Vitamin	0.4	

Turkeys

Energy

As with the chickens during the first week, turkey poults have to be brooded and therefore fed in the vicinity of the brooders, often an paper or cardboard under a high level of lighting. Energy concentration should be not less than 12.6 MJ per kg.

The growing poult requires a high energy diet if growth rate is to be near the bird's genetic potential. This is similar to that provided for broilers, that is about 12.3 MJ per kg. Between 3 weeks and 16 weeks there can be a reduction of energy concentration to 11.0 MJ per kg. During the growing stage some small-scale producers may use whole grain as part of the ration, making up with pelleted grain balancer diet. It is at this stage that hen and cock turkeys (stags) are usually reared separately, since male birds grow faster, are larger and may prevent the hens obtaining sufficient food.

During the final 3–4 weeks before sending for slaughter, a high energy ration is again given but with protein content reduced.

Protein

The overall protein requirement is at the highest for the turkey poult stags (0-8 weeks) when it should be 280 g crude protein per kg, with lysine at 12 g per kg and methionine + cystine at 8 g per kg.

There can be a reduction in protein level at the early and late grower stages (8-12 weeks) and 12-18 weeks to 220 and 190 per kg, respectively; the finishing stage, 18-24 weeks, is further reduced to 160 g per kg. At the present time there does not appear to be reliable estimates of individual amino acid requirements.

Turkey breeders are kept on somewhat lower energy and protein intakes. Eggs laying begins soon after 30 weeks and with some strains there may be a need to reduce intake, as they may otherwise get fat. Protein levels are kept at about 160 g per kg.

Minerals

Turkey poults have been shown to be particularly sensitive to both excess and deficiency of calcium and phosphorus. It is therefore important to see that the level of calcium is between 8 and 9 g per kg and that phosphorus is between 6.5 and 7.5 g per kg. At the growing stage, the level may be reduced to 5 g per kg for each element, while for finishing, 4 g per kg appears to be optimal.

For breeding turkeys being fed at 12.6 MJ per kg, the laying hens need to receive 24 g per kg of calcium and 5 g per kg of total dietary phosphorus. Turkeys, while poults, are sensitive to any excess of salt. Their sodium requirement is only 1.5-2.0 g per kg of feed.

Manganese is a trace element that must be present at a level of 50 mg per kg of perosis is to be avoided.

Vitamins

The requirement for retinol (vitamin A) for poults and growers is 4400 IU per kg, although it is likely that commercial mixed vitamin and mineral concentrate mixtures give higher levels. The cholecalciferol (vitamin D_3) requirement has been set at 900 IU per kg.

There has been an increasing number of reports of biotin deficiency in turkeys during recent years. It is important to see that there is at least 0.25 mg per kg in the diet.

Ducks

The market for ducklings reared to 8 weeks has been expanding as more intense selection and hybridization proceeds. It means that, like broilers, the need is for a high energy concentrated diet with a matching level of crude protein containing balanced amino acids.

A diet containing 16.3 MJ per kg and a crude protein content of 270 g per kg has been found suitable from 0 to 14 days. The later stages of growth and fattening will be satisfactory if the broiler feeding regime is followed.

There is a lack of critical feeding experiments with ducklings to discover the correct levels of minerals and vitamins, but in practice the same level of supplementation given to broilers are found to be satisfactory.

Rabbits

In Europe, particularly in France, there is a strong and growing demand for rabbit meat. Selection is steadily making earlier maturing strains of the New Zealand and Californian White rabbits available. Knowledge of their nutritional requirements is still scanty.

Although a monogastric animal, the rabbit has a large caecum and colon where undigested fibre is fermented and protein degraded. Through coprophagy of the soft faeces, the rabbit obtains additional nourishment, including water-soluble vitamins.

Small-scale or backyard rearing of rabbits for the table is of course still undertaken. These rabbits are often fed a variety of green plants such as surplus vegetables or mixtures of herbs and grasses gathered from nearby pastures, generally supplemented with cereals and kitchen scraps.

The expanding numbers of commercial rabbit producers must feed a diet of constant composition adequately balanced for the animal's needs. It is now the custom to obtain a pelleted mixture from feed manufacturers that contains dried grass or lucerne (alfalfa) meal, ground cereals and oil-seed meals. This provides the necessary 12-18 per cent of fibre with energy at about 11 MJ per kg DM and crude protein between 160 and 170 g per kg DM.

The present belief is that the daily energy requirement of breeding does, ranging from 2.5 to 5.0 kg, are as follows:

Body weight, kg	*Maintenance MJ/day*	*Growth, MJ/day*	*Pregnancy MJ/day*	*Lactation, MJ/day*
2.5	0.895	2.386	1.193	2.585
5.0	1.505	4.018	2.006	4.347

There appears to be a considerable range in the protein levels in commercial mixes, from 14-18 per cent for growers and 16-20 per cent for pregnant and suckling does. The recommended levels for lysine, methionine + cystine and arginine are 6.0, 4.5 and 5.0 g per 100 g protein, respectively.

Table 3.10. Mixture for the rabbit diet

	Composition, %, for			
Constituent	*General use*		*Fattening*	
Barley meal	—	20	—	
Ground oats	12	20	75	—
Maize meal	—	10	—	30
Weatings	18	—	—	105
Bran	40	—	8	
Grass meal	20	20	10	30
White fishmeal	10	—	—	2.5
Meat and bone meal	—	10	—	
Soya-bean meal	—	20	—	12.5
Sunflower meal	—	—	—	4.5
Protein concentrate (used in pig diets)	—	—	—	10.0
Vitamin/mineral supplement at approx. 2.5% (CP = 21%)				

The recommended vitamin levels for caged rabbits are, retinol (vitamin A) 7000 IU, cholecalciferol (vitamin D) 900 IU, a-tocopherol (vitamin E) 40 mg, vitamin K, 2 mg per kg diet.

It is believed that mineral requirements will be satisfied if calcium and phosphorus are kept at 5 g and 4 g per kg of the diet, salt 5 g, magnesium 2 g and manganese 40 mg per kg of diet.

Mixtures that have been found to result in maintaining fertility in does and satisfactory growth rates in their offspring are shown in Table 3.10

Adult bucks and resting does are normally restricted to approximately 113 g per day. The pregnant doe needs 198-226 g towards the end of pregnancy. Lactating does and growers are fed ad lib.

DOGS

A continuing popular misconception over the feeding of dogs is that, since they are classified as carnivores, it is desirable to feed them a wholly meat diet. Under natural conditions, the dog will consume all its prey and this will contain a considerable quantity of carbohydrates and fats as well as protein. Thus it is not necessary to feed a dog more than approximately 22 per cent of its diet as protein. Indeed, dogs have been kept in good health when fed on vegetable protein to which some additional methionine has been added.

There is great variation in the size of dogs and thus their requirement for food. There is also a great deal of individual variation in the way that any dog will respond to the food given it. When a group of 24 English Setters were fed for two years on the same diet, some of them kept their original weight of between 43 and 53 lb while others gained up to 109 lb.

Maintenance Requirements

The National Research Council of the U.S.A. gives the estimated daily food requirements for the maintenance of dogs of different weights.

Table 3.11. Estimated daily food requirements for dogs of various weights

Weight, kg	*Dry-type food, kg/dog*	*Canned or wet-type food, kg/dog*
6.8	0.17	0.43
13.6	0.28	0.72
22.7	0.42	1.06
49.8	0.75	1.90

Dry-type diets are assumed to contain 90 per cent dry matter, 24 per cent protein, 10 per cent fat, 40 per cent starch and sugar and 10 per cent fibre and ash. The canned or wet diets are assumed to contain 75 per cent dry matter, 20 per cent protein, 8 per cent fat, 34 per cent starch and sugar, 8 per cent fibre and 5 per cent propylene glycol.

A useful guide to the daily maintenance requirements of adult dogs is to maintain the proportion of 4.8 g per kg of bodyweight per day for protein, 1.3 g per kg bodyweight per day of fat and 13 g per kg bodyweight per day of carbohydrates. This gives an approximate ratio of 1 part of protein to 4 parts of carbohydrates and fats.

For growing puppies the requirement of protein is somewhat higher at 9.6 g per kg bodyweight per day with the fat at 2.2 g per kg bodyweight per day.

Pregnant Bitch

During the first seven weeks of the bitch's pregnancy, a maintenance diet is all that she needs since it is only during the final two weeks of her pregnancy that the foetuses make most of their growth. During these final weeks the amount of food given to the bitch should be increased by between 20 and 30 per cent. If the bitch is carrying a large litter as one may judge from the size of the abdomen, it is as well during the final few days of her pregnancy to give a more concentrated diet of milk and cheese and eggs, since the pressure on the stomach and intestines from the enlarged uterus will interfere with the digestion of foods which contain a proportion of fibre.

Feeding during Lactation

The milk of the bitch contains more than twice as much protein as does cow's milk (7.6 vs 3.5 per cent) and more than twice as much fat (11.0 vs 4 per cent); thus it is important for the bitch to be given additional protein and additional energy in her diet. She should in fact be fed at approximately three times the maintenance level.

Since the bitch will be losing a considerable amount of calcium in the milk, it is necessary to be sure that she has additional mineral supplements and it is as well to add 10-15 g of steamed bone-flour to her daily intake of food.

Feeding Puppies

For the first three weeks, the puppies will be consuming the bitch's milk alone but by the 4th week they should be consuming some solid food as well. They can be given some small pieces of meat.

Cats

It is necessary to be clear about some physiological facts concerning cats if feeding is to be successful. Even more than the dog, they need a high intake of animal protein and they digest fat well. Wheat cereal foods such as bread and unsweetened biscuit or cooked potato can be readily digested, but uncooked starch, lactose and sucrose should not be given. Cats are a desert-type animal that in the wild depend largely upon water (about 70 per cent) in the body of its prey, only occasionally drinking water. Thus, dry cat foods, even when water is separately available, may cause urolithiasis or the deposition of calculi in the urinary tract.

Cats are creatures of habit and unless early trained to accept a varied diet, may firmly refuse to eat different foods. So long as the established diet is one correctly balanced, not to have to seek new assortments of foods can be convenience.

Adult Cats

The minimum *protein* content of the diet for an adult cat is 21 per cent, but it is better to keep it at 25-30 per cent. Doubt has recently been thrown on this long-held belief by work on the cat's amino acid needs. Through the use of purified amino acid diets it has been shown in the U.S.A. that so long as these are in the correct concentrations, maximum growth could be obtained in kittens by only 16 per cent in such a diet. Burger et al. (1981), by making adjustments to the amino acid profile of diets containing different protein levels, were able to show that adult cats could be kept in nitrogen equilibrium with the dietary protein at only 12 per cent. The accepted belief has been that this protein must be meat or fish to provide the range of amino acids necessary. Any type of muscle meat is suitable, as also is liver, but lung tissue is not suitable as an only source. Meat is best fed raw, so long as it fresh and bacteriologically uncontaminated. Some cats prefer meat cooked, which is nutritionally satisfactory so long as it is not over done. Some commercially prepared tinned meats that may have been heated for too long or at too high temperature have some of the protein denatured, forming N-glycosides, which are not digested.

The *fat* content of most commercial cat diets is between two and eight per cent, of which one per cent should consist of the unsaturated linoleic or arachidonic acids. A high level of fat can safely be added, for in the wild a cat obtain 60 per cent of its energy from fat.

Carbohydrates can then be added to the extent that the total energy of the daily food requirement reaches the following levels:

	Weight, kg	*Daily energy intake, MJ, per kg*	*Daily ration, g*
Neuter (male)	4.0	0.34	200
Neuter (female)	2.5	0.34	140
Male (tom)	4.5	0.34	240
Female (queen) pregnant	3.5	0.42	240
Female (queen) lactating	2.5	1.05	415

It is during the last 2-3 weeks of *pregnancy* that both quality and quantity of the ration should increase. Now the protein should be at 30 per cent or more and fat 10 per cent or more and the amount given daily should be increased by 25 per cent.

Kittens begin suckling while they are still being licked dry, and so long as they receive sufficient milk, will gain about 10 g per day. With 12 feeds in 24 hours and perhaps for kittens to suckle, it is clear that the *lactating female* must be given all the high quality food that she can eat. It should in fact be at least 2.5 times the amount she was eating at the beginning of her pregnancy.

Young Cats

By the age of 4 weeks the kitten must be given some solid food as well as its dam's milk. Each kitten should be given one small meal made by mixing milk with cooked oatmeal, Farex, or bread, and one or two small meals of minced beef or liver (raw or lightly cooked), or finely cut-up cooked fish. Half a teaspoon of feeding (sterilized) bone-meal should be given with one of the meal meals each day, and once weekly a half teaspoon of cod (or halibut) liver oil.

After weaning at 6-8 weeks, the weaners should be given a main meal of 1 part of animal protein, such as boiled fish, or cooked liver or tinned meat, with 2 parts of boiled mashed potatoes and 1/200 parts of dried yeast. Another meal can be made of 1 part of Farex or biscuit meal, 1 part of full cream milk powder made into a stiff paste, for cats dislike sloppy foods. They will be eating about 200 g per day by 10 weeks, which should contain at least 33 per cent protein and 10 per cent fat.

If the growing cat is given 100 ml of milk daily, it will provide 125 mg of calcium, but its requirement is for between 200 and 400 mg per day, so it is advisable to continue to give small amounts of bone-flour until bones themselves are given and eaten.

Since cats are unable to convert carotenes into retinol (vitamin A), it is particularly important to see that growing cats are given a supply of the vitamin, either through regular meals of liver (3 × per week) or by giving it as cod (or halibut) liver oil. Reputable commercial prepared cat foods will have the vitamin added.

The rearing of orphan kittens can be satisfactorily achieved if it is remembered that cat's milk has twice the amount of protein and nearly twice the amount of fat that is contained in cow's milk. A substitute milk can be made by dissolving 20 g skim milk powder in 90 ml of warm water, then adding 10 ml of warm vegetable oil and feeding the mixture at 38°C. The oil should contain 80 μg per ml of retinol (vitamin A) for the first two days, then 50 μg per ml is sufficient. At first the kitten will only take 2 g per feed and it must be fed 12 times during the day and night. The amount fed is gradually increased to 10 g per feed, giving 7 feeds in 24 hours by the age of 4 weeks, when they should be lapping the milk from a saucer. In the early days, an eye-dropper, or glasstube with a rubber bulb at one end, or a plastic hypodermic syringe may be used.

WATER

Although the importance of providing a plentiful supply of pure water for all types of livestock is easily admitted, it may not always be understood what this involves. It is easy to see that some species, sheep for example, especially the fat-tailed breeds, zebu cattle or camels can survive on very meagre and uncertain water intakes. This they can do because each of these species has evolved in regions of high summer temperatures and frequent, sometimes prolonged, periods of drought. They can extend survival through the use of physiological mechanisms that permit concentration of urine and the release of body water held in tissue reserves. On the other hand, a high-producing dairy cow at the peak of lactation must have ready access to a large volume of water for several times during the day if her production is to be maintained.

It is clear that animals grazing young grass or fodder crops which may contain 90 per cent of water can obtain almost the whole of their daily requirement of water in this manner. Those types of ruminants adapted to the tropics, possessing fat storage appendages, can make use of the mechanism of oxidation of these reserves to provide needed water, which of course is available to any animal in an emergency. It is calculated that oxidation of 1 kg of fat will provide 1070 g of water. Avoidance of this wasteful use of body resources rests with the owner to see that there are supplies of safe water available whenever livestock wish to drink.

The amount that will be drunk depends upon two main factors. First, there is the individual's particular needs, for these can differ as much as 50 per cent in cows being fed the same ration. Secondly, there is the dry matter content of the food being eaten. For cattle, it may vary between a ration of straw and concentrate with 10 per cent moisture, to roots or a green fodder crop already mentioned.

Water Requirements

Horses

The water requirement is 36-45 1 per head per day, but this is increased by sweating, work or dry food and high temperatures. It has always been held that excessive drinking after exercise can lead to indigestion (colic). It is probably wise to allow tired horses to drink only to a partial fill before resting, with more given later. Stabled horses should have a constant supply of water in their drinking bowls.

Cattle

The needs of calves will vary with the temperature and dry matter content of the ration, but a guide is 4.61 per 45-kg body weight.

Cows in milk have a varying need according to the amount of milk being produced. A ratio of 5 : 1 of water to milk is stated. A dry cow needs 36.5-451 per day, increasing to 701 per day during the last part of pregnancy, It may also be calculated on the basis of dry matter intake. In the thermoneutral zone 15-20°C. it is 4.11 per kg DMI with 0.871 per kg milk produced.

Sheep

Adults need 4.51 per head per day, with half this amount for lambs when being given dry food. Pastures in temperate climates, with the addition of rain and dew often provide enough water. Pregnant ewes need up to 2.2 times that for dry ewes. Lactating ewes need more water, calculated as 1.5 times that for dry sheep.

Pigs

Only if liquid feeding or whey is being fed ad lib will it be unnecessary to provide water to pigs. Weaners: 3.41 per head per day, increasing to 5.7-6.81 at bacon weight. Pregnant sows and boars: 181 per head per day. Lactating sows: up to 231 per head per day. Self filling water bowls or nipple drinkers are most commonly used, placed away from bedding.

Rabbits

The use of nipple drinkers is widely adopted. Doe (dry) : 0.281 per head per day, increasing to 0.571 in pregnancy. Lactating doe: 3.41 per head per day. Nipple drinkers must be tested daily.

Poultry

Broilers and growers kept on litter must have sufficient water bowls spread evenly, 1 : 50 birds to allow ease of access.

Dogs and cats

It is important to see that water is always available. Although cats are desert animals, if given dry food, they must be encouraged to drink to avoid urolithiasis.

4

BREEDING

For many centuries the breeding of domestic livestock has been wholly based upon the simple fact that visible characteristics in the parents are passed on to their offspring. Once a farmer identifies what he believes to be desirable characters, amongst which are obvious things like colour, shape or size, he selects for them. He will search for dams and sires possessing these particular characteristics, knowing that the offspring are likely to have them also. It would have been in this way that over many centuries men produced horses possessing an increased capacity to carry a rider with greater speed or to pull a heavier load, because horses were of particular value to early organized societies and they were kept under close control. Selection for desirable characters would have proceeded with various farm animals in a similar fashion.

HISTORICAL OUTLINE: THE ANIMAL IMPROVERS

As the industrial revolution developed in western European countries in the 18th century, there was a clear need to increase the supply of meat and milk and other animal products. This was the challenge that the 'animal improvers' answered to such good effect that the foundations were quickly laid for a number of modern breeds of livestock. One of the best known of the men who sought to meet the need of the period was Robert Bakewell of Dishley Grange in Leicestershire. The way that he produced the New Leicester breed of sheep has been relatively well documented. In his time, 1725—1795, the roads connecting various towns in England, as in other European countries, made travel difficult. Few farmers travelled beyond their market town and there was little interchange of breeding stock from one part of the country to another.

Livestock had remained relatively isolated for centuries and, through adaptation if not active selection, developed different characters of colour, size, rate of growth, or capacity to undertake draught (with cattle) or produce wool and milk (with sheep).

The accomplishment of the 'Improvers' like Bakewell was that they deliberately sought to produce heavier sheep and cattle, with early maturing carcasses. The device of mating closely related individuals known as inbreeding, with the hope and expectation by this means of more quickly enhancing a desirable character was widely used by Bakewell. He, like other breeders of the time, collected together a very mixed group of animals, i.e. their appearance or phenotype was variable, as was bound to be found in their neighbourhood. They thus began with animals of a wide genetic base or genotype. By close breeding and ruthless culling of 'off-types', Bakewell achieved a rapid improvement in the particular characters for which he selected and, therefore, a diminution of genotypic and phenotypic variability. The method of inbreeding concentrates good and bad characters without distinction and, if continued for too long, inevitably results in diminished fertility. By hiring out to his neighbours a number of his better rams in order to see the sort of lambs they sired, he began a type of progeny testing, a rather more elaborate form of which is used extensively today for discovering superior sires in most breeds of farm livestock.

The effect of the rapid application of such methods, up and down the country, laid the foundation of all modern breeds of cattle, sheep and pigs. Not only did the New Leicester itself represent an important advance in practical farm livestock, but, following the work of Culley in Northumberland in crossing it with the Cheviot, the 'Border Leicester' was produced, so important in producing the popular Half-bred type ewe of today. Likewise, the manner in which one of Bakewell's former pupils, Charles Colling from Northumberland, adopted these principles, along with his brother Robert, in the formation of the 'Shorthorn' breed of cattle was similar.

Emergence of Modern Breeds

Nearly all the European breeds of livestock became fully delineated as separate entities during the 19th century. Quite large numbers of some of the more popular breeds such as the beef and dairy Shorthorns, Ayrshire and Hereford, were exported to the U.S.A., Canada, Latin America, South Africa, Australia and New Zealand. While several other breeds came from parts of the continent, by far the greatest number were sent from the British Isles. Adaptation to new needs and different environment effects in the American and Australian continents have produced many new breeds of farm livestock.

There was, during the nineteenth century, great improvement in the means of public transport through the new railway networks and better roads. This encouraged the establishment of agricultural shows which proved an enormous stimulus for the breeders of the locally numerous types of livestock to form *breed societies* and exhibit their registered stock competitively year by year. The breed societies made accurate recording of pedigree animals obligatory, but there was no more science attached to breeding farm livestock than in Bakewell's day. Throughout the 19th century and into the 20th century breed societies multiplied and flourished, based upon a buoyant demand for pedigree animals, especially those known to have won awards at local and national shows.

Scientific Basis of Stock Breeding

Knowledge obtained by the use of steadily improving histological techniques, with new efforts to reveal the differing substances found in the cells that make up the organs of the body and with the availability of better light and electron microscopes, permitted remarkable progress towards an understanding of the nature and purpose of the differing parts of the individual cell. First, the chromosomes which carry the genetic information were revealed and the way in which various body cells multiply; after this the method of reduction division in the developing sex cells of the male and female was revealed.

The concept of the *gene* as the ultimate factor determining the shape and nature of individual animals was put forward many years ago, but it was not until 1952 that its molecular structure, the double-coiled strands of deoxyribonucleic acid (DNA), was revealed. On this is encoded all the organism's genetic information. Subsequent work showed how this information is passed to the *ribosomes*, bodies within the cytoplasm of the cell, where ribosomal transfer ribonucleic acid (RNA) transports the amino acids to be subsequently built-up into proteins, after the nucleotides peeled off from the DNA strands in the nucleus are brought by messenger ribonucleic acid (mRNA) to permit the synthesis of proteins. It is these, in the forms of enzymes and hormones, as well as structural proteins, that determine the shape and function of tissues.

Cell Genetics

Every species of animal has a particular number of chromosomes in the nucleus of every body cell (known as the 'diploid'), with the exception of the germinal cells in the sex organs. In the process of the division of all these body or somatic cells as tissues grow or renew themselves, the chromosomes with all their genes go through a process of condensation, alignment and equal division: a process known as *mitosis*. This diagrammatically for an animal with only four pairs of chromosomes. Higher, and thus more complex animals have many

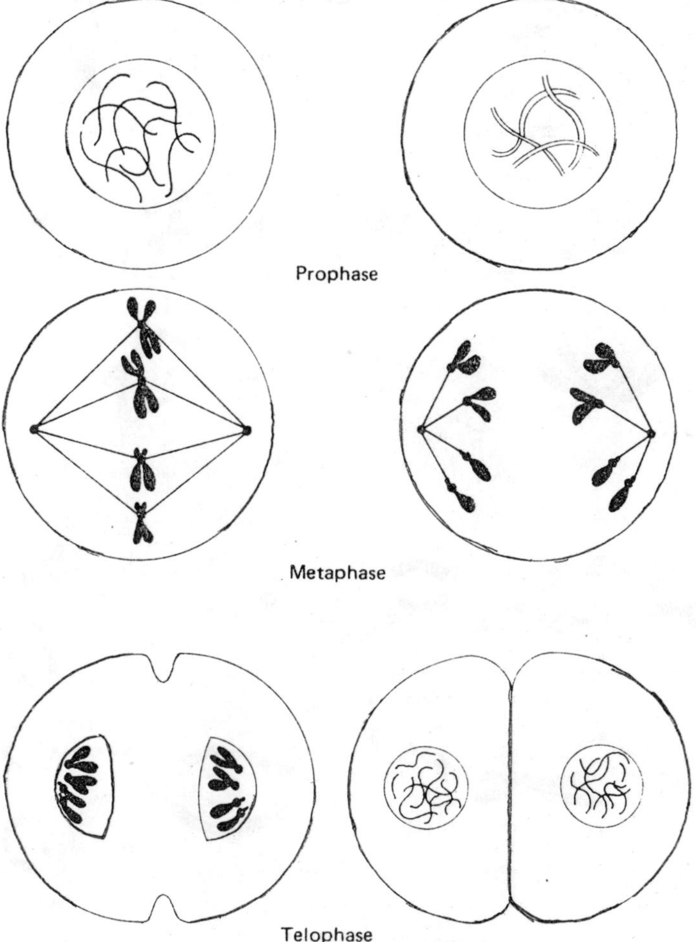

Fig. 4.1. Mitosis in cell with four pairs of chromosomes.

more chromosomes, e.g. cattle have 29 pairs, horses 64 pairs, sheep 28 pairs, pigs 38 pairs, dogs 78 pairs and cats 38 pairs.

A moment's thought make one realize that when mating occurs and the gametes of male and female, with their equal numbers of genes, combine to produce offspring, there must be a mechanism to

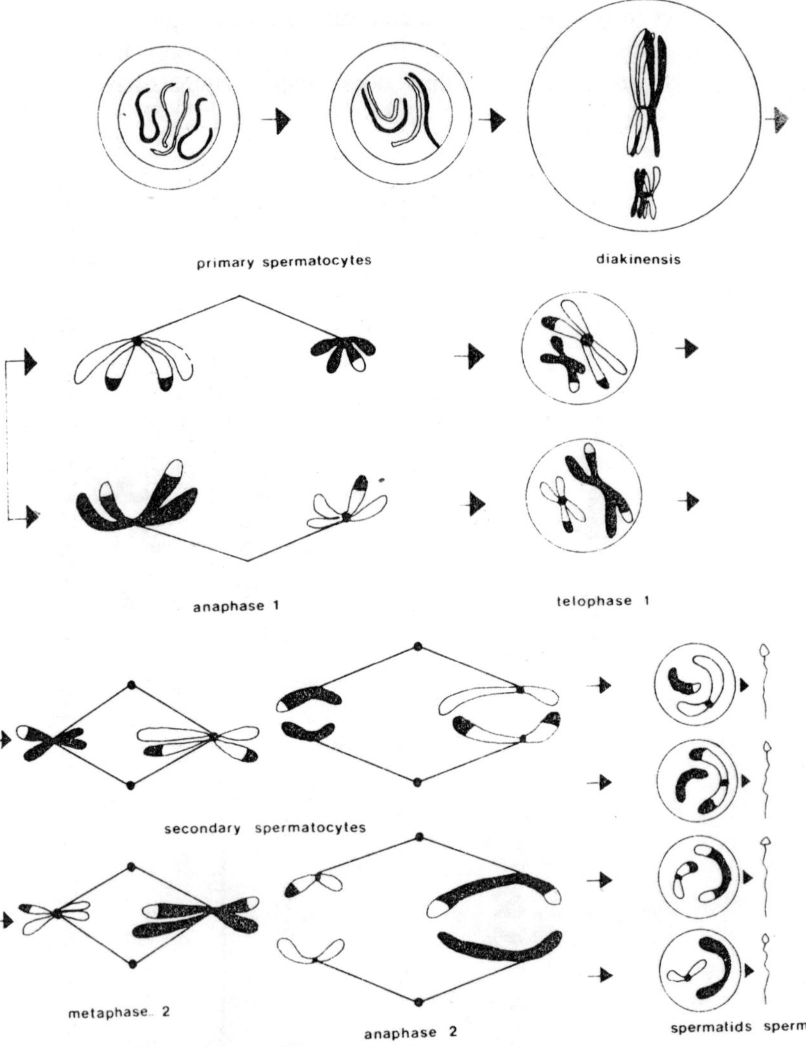

Fig. 4.2. Steps in meiosis in one animal with two pairs of chromosomes, one pair (white) from dam and one pair (black) from sire.

halve the number of chromosomes, so that the correct number may be passed on to the new individual, the zygote.

This reduction division or meiosis occurs in the developing sex cells, that is as the ovum of the female develops in the ovary and the sperm develops in the testis of the male.

It is not very surprising that things go amiss during this more complicated type of cell division and chromosomal abnormalities result. Occasionally, there can be alteration in the number of chromosomes at the more numerous mitotic cell division. When the number of chromosomes change so that there are either one more or less, that is $2n + 1$ or $2n - 1$, where n, is the haploid number, there may be profound changes in a developing embryo, making it impossible for it to survive. It may be aborted or it may be resorbed if it is at a very early stage of growth. In either instance, an animal thought to be pregnant returns for remating with consequent loss of valuable time. Sometimes there are inversions, duplications and translocations of the chromosome parts, most causing small alterations from the normal but possibly revealing itself at some stage after birth. With the development of techniques for culturing and staining cell contents to allow examination of chromosome structure (karyotyping), a clearer picture of the extent and significance of these changes becomes available.

Mendelism

The rapidly extending knowledge of cell chemistry has helped to explain the empirical findings of the earlier breeders based upon. Mendelian genetics. It was the great achievement of Gregor Mendel, embodied in his publication in 1866 in which he gave the results of crossing various types of sweet peas, that he appreciated the fact that characters or traits were passed on as discrete units and that many were dominant. He showed that in subsequent interbreeding, the dominant and recessive alleles (those characters linked on the same pairs of chromosomes) separated in a constant manner, depending upon the number of dominant and recessive characters in the organism being studied.

When his work became generally known during the early years of this century, animal breeders were able to show that dominant and recessive alleles existed with all common breeds of livestock. They control the appearance of a few characters of economic importance, such as horns in cattle, coat colour in cattle, comb shape in poultry.

The term 'homozygous' is applied to an organism that has the character under investigation in pure form, that is both of the alleles are present, either in the dominant or the recessive form. When homozygous individuals mate, the progeny will possess one of each allele. This is the heterozygote, and every heterozygous individual will appear exactly similar to the dominant homozygote parent. Thus, all the progeny from horned and polled parents will be polled, because 'polledness' is a dominant character.

Segregation occurs in the following generation if the heterozygotes are themselves mated. With a monohybrid cross involving one dominant character, the offspring in the second (F_2) generation will reveal the character in the ratio of three polled calves to one horned calf. Only one of the former calves would be homozygous for the polled character and thus able always to transmit hornless genes, since the other two will carry the recessive horn gene. There has to be another mating amongst heterozygous animals to discover which one is the homozygous dominant.

It was soon discovered that some characters were not inherited in a simple Mendelian manner. Such a character is the red coat colour of Shorthorn cattle, with white hair as the other allele. The progeny of matings between red and white Shorthorns are neither red nor white, but a combination of red and white hairs producing a roan coat colour. When roan cattle are mated the progeny are likely to be in the proportions of: red: 2 roan: 1 white.

Table 4.1. Inheritance of the Shorthorn's coat colour

	Red bull	×	White cow	
Parents (P)	RR		rr	
gametes	R	Roan	r	
Offspring (F_1)	Roan	Rr	Roan	
$F_1 \times F_1$	Rr	×	Rr	
gametes	R r		R r	
Offspring (F_2)	R		r	
R	RR red		Rr roan	
r	Rr roan		rr white	1 red: 2 roan: 1 white

Most of the characters which we want to improve in any domesticated species are the result of the action of more than one gene and they may have their effect by various mechanisms. By the employment of a variety of techniques, geneticists have been able to gain a clearer picture of how these characters are inherited.

Effect of the Environment

What is always so difficult to measure when we are trying to increase a desirable trait is the extent of the influence that the environment is exerting. The environment has to be seen as the sum of all the factors that affect the multiplication and renewal of the cells of the individual animal. The environment of the uterus, the nourishment the developing embryo receives, and the later more obvious effects of housing, climate, the care given by attendants, and the effects of disease, all combine to make up the total environment.

This interaction of the inheritance of an animal, its genotype with its environment, 'the genotype × environmental interaction', as it is called, can cause difficulties when animals or their offspring have been selected in one place and then taken into very different surroundings. It has been found that not only may there be an alteration in the overall performance, as happens for instance, when European cattle are transported to the tropics, there is also the possibility that the ranking or expression of certain characters may be changed by particular factors, such as the temperature, in new surroundings.

Dairy Cattle in the Tropics

An illustration of this is provided by the effect of high temperatures upon the capacity of the Friesian cow to produce milk in the tropics. In temperate surroundings, the average production of Friesians today is within the range 3000-4000 kg in a lactation of 305 days. Genetically similar cows, when placed in many tropical countries, are hard put to it to produce 1000 kg in a lactation.

High temperatures have a direct effect upon most European breeds of cattle, causing symptoms of heat stress, a drop in milk yield, lack of appetite, which itself contributes to lowered production. It is sometimes stated that the upper critical temperature for European breeds of dairy cows ranges between 21-25°C. It is apparent that in the tropics they become adapted to higher temperatures. Individuals that are unable to adapt are culled within the first two years of their arrival.

Both the Jersey and the Brown Swiss adapt better than the Friesians to continued high temperatures. Although overall production may be reduced, there is often not the steep fall during the hottest months. This is the reason why these breeds have been so widely used in the tropics, often crossed with zebu dairy breeds, such as the Sahiwal, Hariana and Red Sindhi

Australia

The remarkable manner in which the Merino sheep, bred originally in Spain as a fine-wool type and hardly exported at all until the 18th century, adapted and thrived in the Australian environment during the last century is well known. In the 1880s a group of farmers in the South Island of New Zealand combined the wool characters of the Merino with the larger frame and better maturing qualities of the Lincoln to produce the Corriedale, a breed well stabilized and used in many countries. More recent stabilized cross-breeds are described later.

U.S.A.

In the U.S.A, the requirement in the Southern States for cattle that could withstand the summer heat and more efficiently utilize the dry pasture, was met by incorporating a proportion, generally three-eighths of zebu blood with the European beef breeds. The breeds of cattle that were gradually selected within the tropical areas of the Indian subcontinent and in parts of Africa, known as *zebus* (*Bos indicus*), are mostly smaller than European breeds. They characteristically have a thinner hide, more and better developed sweat glands, a prominent dewlap (a fold of skin from chin to brisket) and often a hump of fat and muscle tissue over the withers. Representatives of several Indian and Pakistani breeds, Angole, Krankrej, Gir, Sahiwal, were brought into the southern United States, and it was some of these that were combined to form what has become known as the 'Brahman'. Thus the breeds 'Beefmaster', 'Charbay', 'Brangus' and others were created as well as the earlier 'Santa Gertrudis'.

In other words, there can be no doubt that the only safe way of selecting for superior performance in any breed or type of livestock is to undertake the selection programme in an environment as closely similar to that existing on farms where the progeny will be kept.

Breeding Programmes

Methods used to Produce Improved Strains of Animals

The traditional methods employed by many generations of progressive livestock breeders to produce progeny that were superior to their parents are obviously still available, but not now so frequently used. They involved judgement by the eye of the breeder and the very extensive use of certain animals, often males, who exhibited the desired qualities to an exceptional degree. Concentration on the conformation of the male as the instrument of effecting improvement is due to the fact that they have more offspring each year than a female. These were often mated to close relatives, known as *inbreeding*.

Inbreeding

This procedure inevitably means that the variability of all characters is reduced, faults or undesirable recessive characters that are not seen in the mixed heterozygous state are now likely to be revealed. This state of affairs is sometimes spoken of as 'inbreeding depression' which some believe is bound to appear as soon as the degree of inbreeding reaches the level of half-sibs or closer relatives being mated. A notable feature is the decline of traits of fitness, i.e. characteristics such as fertility and mothering ability, while embryonic mortality increases in females and libido decreases in males. These adverse effects are due to several pairs of recessive genes, each having a slight detrimental effect on some character, becoming reinforced in their manifestation in the inbred offspring. Demonstrable action is the result of failure of these genes to produce the necessary enzyme for a correct cell action, or is through the production of an abnormal protein.

The way that continued inbreeding can have practical deleterious effects can be seen in such recorded episodes as the line of pigs at the Missouri Experimental Station that had to be discarded because the recessive gene for haemophilia caused many to bleed to death at castration or at farrowing. With cattle, it was shown that when inbred cows were inseminated with semen from related bulls, there was much greater embryonic mortality, with only 36.8 per cent completing their pregnancies, compared with 65.7 per cent with non-inbred animals.

During the last century, Thomas Bates, a pioneer breeder of the Dairy Shorthorn, used inbreeding to reduce variability in milk production, but in so doing, he lowered the fertility of the 'Duchess' line of cows to such an extent that, of 58 cows, 24 were barren. Milk production will also be reduced if inbreeding is persisted in, to the extent that for every 10 per cent increase in the coefficient of inbreeding there is likely to be a reduction of 6.6 kg, or 5.2 per cent, of the non-inbred mean.

If close inbreeding is not continued for more than a few crosses and if severe culling of all animals showing any sign of recessive characters is undertaken, it is a fact that the increased homozygosity of parents confers a high degree of prepotency. This term simply means that offspring will be very similar to the more homozygous parent.

It has already been stated how the early 'Improvers' of the 18th century used these methods. They have occasionally been used by those creators of new breeds of the 20th century, like the Klebergs of King Ranch, Texas, with their Santa Gertrudis breed of beef cattle, three-eighths Brahman and five-eighths Beef Shorthorn, or Colburn of Northleach, in England, who produced the 'Colbred' sheep from a combination of Clun, Border Leicester, East Friesland and Dorset Horn breeds.

Mostly, these and many others have used a less intensive form of inbreeding known as *line-breeding*. This means that mating is planned to maintain a close relationship to an unusually desirable individual, often a sire possessing special qualities. It was due to line breeding and strict culling that the useful American quarter horse was produced. The name derives from its capacity to stop and turn sharply, so proving most useful in cutting-out or quartering individual cattle amongst a mob when yarded.

Mass Selection

If it is desired to improve some character in a species of farm animal, often within a particular breed, it is important to be able to measure it, either by direct measurement or by measurement of a linked character, such as food conversion which is linked to the rate of weight gain. The commonly used method of selection is to use as parents only those males and females that are superior to the average. The method is known as individual or mass selection, for it is the result of measuring characters in a series of individuals. The greater the number of animals included and the smaller the variation from the mean, the more effective the result will be.

How much of the improvement over several generations is due to the success of selecting superior gene combinations and how much to improvement in the environment, such as better control of disease, better stockmanship, improved nutrition, cannot be measured unless equal negative selection has simultaneously been undertaken, or a randomly mated control population has been maintained that consists of equal numbers of males and females as the selected population. The expense associated with either device means that it is only Government agencies or large commercial companies that can undertake such schemes.

In the United Kingdom, such a body is the Animal Breeding Research Organization, an institute under the control of the Agricultural Research Council. At its several farms, it has undertaken the assessment of newly imported breeds of farm animals and, later, proceeded to incorporate any especially beneficial genes they may possess by crossing them with selected British breeds.

It was in 1975 that they announced a new breed, the *Damline*. Table 4.2. shows those breeds that McClelland (1975) stated were selected for inclusion in the gene pool from which the new breed was synthesized. After a number of years of crossing, selection began in 1972. Ram lambs are selected on the basis of the 8-week litter weights of their dams and other female relatives, and ewe replacements are initially selected on the same basis and are culled or kept on their own performance. Other selection criteria are such traits as fertility and lamb mortality, as well as growth rate. When Damline rams were mated to Scottish Blackface ewes and their offspring compared with Border Leicester × Scottish Blackface lambs kept under the same conditions, clear superiority was shown based on numbers of lambs weaned and 8-week lamb weights.

Table 4.2. Breeds selected for inclusion in gene pool for production of the new breed Damline

Breed	*Desirable character*	*Inclusion in gene pool, %*
Finnish Landrace	Prolificity and early sexual maturity	50
East Friesland	Milk production and size	21
Border Leicester	Growth rate	17
Dorset Horn	Length of breeding season and good conformation	12

More recently, claims have been made for the *Cambridge* breed as a producer of useful sires for crossbreeding. The breed consists of 60 per cent Clun Forest, 25 per cent Finnish Landrace and 15 per cent other breeds (apparently as many as 10). The average lambing percentage in pure-bred flocks is 280 and it is claimed that half-bred flocks show high prolificity.

Two new breeds have been produced in New Zealand, both claimed to be superior to either the Romney or Corriedale who in the past have been most widely used as dams for the production of fat lambs. A selection of crosses of the Cheviot and the Romney has resulted in the creation of the *Perendale*, found to be very suitable in the harder hill country. In the South Island, a fixed half-bred between the Border Leicester and the Romney was named the *Coopworth*, and again it was developed to give greater prolificity and high live-weight at weaning.

In Australia, where over 90 per cent of the sheep population had been developed for wool production, a project was started in 1965 to develop a dual-purpose breed with the emphasis upon meat production. This was achieved by mating high quality Corriedale ewes with Border Leicester rams born as twins or triplets from a number of stud flocks. This is now a fixed half-bred and has been given the name of *Gromark*. There are two greatly used devices that can help to identify superior males.

Performance testing

It is recognized that characters having a high heritability, which is a measure varying between 0 and 1 stating the extent of the likelihood of any particular trait being passed on to the next generation, are those that can be measured with relative ease. These include, rate of weight gain, with its associated food conversion efficiency. Under the conditions of optimal nutrition, it has been observed that those animals with the fastest weight gains also utilize their feed most efficiently. In cattle, from the results of measurements in many breeds in many parts of the world, the heritability of final weight is found to be as high as 0.7.

The superior beef bulls most likely to transmit this trait can be identified by measurement of their weight gain over a certain period while they are still calves, and this can also be used with boars. It is important to keep all animals under test in the same environment, including diet, management and the animal attendant, as well as the housing. Those with the highest gain in the stated period are those that should be used for breeding.

The widespread use of artificial insemination in the cattle industry means that most of the male calves produced by dairy cows must be sold. With present methods of semen dilution, storage and transport, it is possible for one bull to inseminate 20,000 cows in one year. Throughout the temperate world, the Friesian or Holstein breed has become the commonest milking cow. Since so many of the male calves are fattened for beef, it is not surprising to find that some countries are superimposing a performance test on bull calves to find those with good 'beefing' qualities. Young bulls, initially selected because of the high level of production of their dams, are commonly tested, as soon as they reach maturity, for their milk producing potential by means of the progeny test.

Progeny test

Characters that are not of a high heritability are often associated with reproduction. It is important, therefore, to measure milk production of as large a number of the daughters of a potential bull, who may be the off-spring of a tested bull and a high producing dam. In order to minimize the effects of the environment, the production of the daughters of the bull under test must be measured in herds with varying patterns of management, and there must be first lactation measurement of a sufficient number of daughters. In many countries, the official progeny test programme specifies a minimum of 20 daughters.

Their milk production is compared to that of the daughters of other bulls in the same herd and of the same age. This type of progeny test is known as the *comtemporary comparison test* and was first used in U.K. in 1953.

From 1974, greater accuracy was introduced through making an adjustment in the sires of those contemporary calves used for comparison. This became necessary because of the present widespread use of bulls of increased genetic merit. From 1979, a further improvement was made by taking into account the merit of the dam's sire, i.e. the maternal grandsire. This test is now known as the *improved contemporary comparison.*

There are similar tests being used in all the advanced dairying countries, most of which are related to a selection index centered upon the calf's dam and her sire. The use of computers has enabled iterative reviews of up-dated results of subsequent lactations to be quickly incorporated into the bull's record, thus giving greater reliability to the choice of bulls for an artificial insemination service.

Widespread use in a region where there are few dairy bulls of great merit from the milk quality and quantity point of view, is likely to have the effect of increasing the closeness of the relationship between calves, i.e. a tendency towards greater homozygosity. This is likely to stop short of inbreeding, but it must inevitably increase the likelihood of undersirable recessive genes, whether lethals or semi-lethals, becoming manifest.

Recessive Characters

Many characters which we would like to be increased, such as more rapid growth, more milk, more fat or solids-not-fat, are measurable. These are sometimes linked to other desirable traits, but occasionally the result of gene action is the manifestation of defects that can result in the death of the embryo or in a foetal monstrosity. Such recessive characters seldom reveal themselves in the mixture of chromosomes that make up a normal heterozygous individual, but they may reveal themselves if a widely used bull happens to be carrying such defective genes.

They will be revealed more certainly and quickly if there is a cross back to some of the close female relatives. This *back-crossing* increases homozygosity, thus revealing the recessive genes.

It is clearly desirable to make some test matings before a bull is put to general use, if the pedigree gives any suggestion of danger. At least it has now become routine to examine the bull's *karyotype*, i.e. the display of chromosomes belonging to each individual which can be made visible by appropriate staining methods, photographing the dividing cells in metaphase, and arranging the cut-out pairs of chromosomes in descending order of size. This shows whether there are any abnormal chromosomes.

Selection index

Another method of selection that is sometimes used in a breeding programme is the selection index. This has already been mentioned as a means of ranking cows who are the dams of bulls undergoing progeny testing. But it can also be a method applied to assess the breeding value of any animal when two or more characters are being simultaneously selected for. The manner in which appropriate weighting is to be given to such factors as genetic correlation, heritability or economic value of the trait, must be carefully considered.

It is generally accepted that the larger the number of characters incorporated into an index, the more accurate it becomes. The conventional method of constructing a selection index is to use the square root of the sum of the traits to be included; thus $\sqrt{2}$ or 1.41 if only two characters are used. Dalton (1980) points out that poultry breeders have included up to 16 characters ($\sqrt{16} = 40$) and pig breeders have used indexes containing up to nine traits. The widely used New Zealand National Flock Recording Scheme called 'Sheeplan' embraces four characters, the number of lambs born, their weaning weight, the liveweight at the hogget stage and the weight of fleece at this stage. Each of these has a predicted average genetic gain ascribed to it and the relative economic value is calculated. The widespread availability of programmable computers has made the calculation of such indices simpler.

Independent culling methods

A rather less accurate method of selection is by using independent culling levels. This means that those individuals who reach a given level of excellence in the exhibition of two or more characters are selected as parents of the next generation.

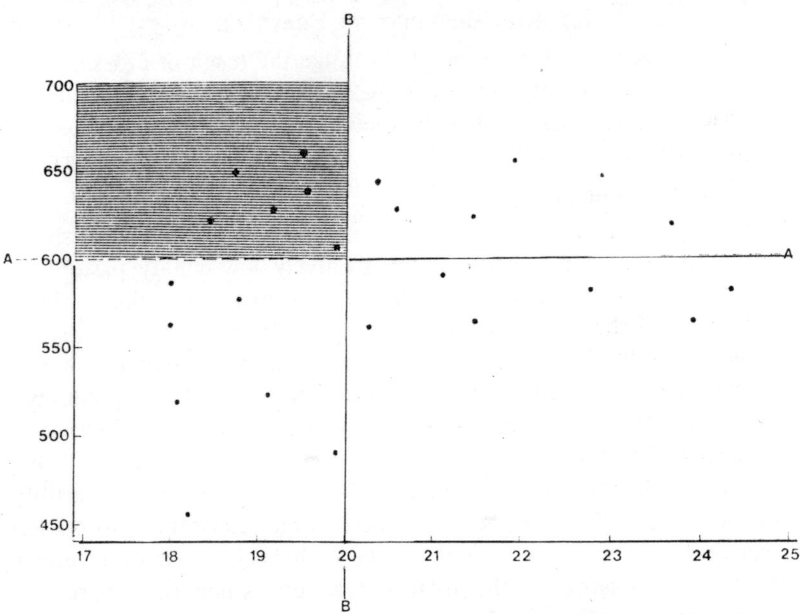

Fig. 4.3. Growth rate and depth of midback fat individual pigs. The culling levels for the two characters are shown as line A–A and B–B. Those individuals whose values are in the shadded area will be selected as parents.

The establishment of what is the appropriate level for selection can be steadily raised, so long as the numbers from which selection is made remain large.

Although it is possible to select for three or even four characters, it means that far fewer animals can meet all the levels set. Progress towards any one objective is slowed proportionately as the number aimed at is increased.

Tandem selection

This is also a somewhat slow and uncertain method of breeding, but has been used with poultry. After one character has been selected for over several generations, selection begins for a second character. There may of course be some degree of negative correlation between the two characters, in which event progress would indeed be slow.

Reciprocal recurrent selection

This is a method that has been used in selecting desired lines of poultry and pigs and has been used to improve strains within breeds, or to test crosses between breeds.

Cross-Breeding in Practice

Practical farmers have long known that the result of crossing two breeds is likely to produce offspring that possess qualities that are superior to either parent. This is known as an illustration of positive *heterosis* and has been extensively used in beef and fat-lamb production.

Fat-Lamb Production

In several countries of the temperate world, a hardy type of sheep has been bred to make use of the relatively low-quality pasture or forage type to be found in the hilly and mountainous regions. These hill breeds of sheep are also adapted to the often harsh winter weather. Although hardy, these ewes are of small size and, under such conditions, of only moderate fertility. When crossed with a ram of a larger breed, reared under a better environment, such as the Border Leicester, Lincoln or Romney Marsh, the half-bred offspring combine the hardiness and vigour of the hill breed with the larger size and greater fertility of the lowland breed. In addition, there is the reasonable expectation that there will be shown some degree of hybrid vigour or heterosis from the crossing of the different breeds since the degree of heterozygosity is increased.

In order to produce the type of lamb that is slaughtered at an early age, usually from approximately 16-weeks old for the well-nourished single lambs desirable to the meat trade, a further crossing with a Downs-type breed, such as Suffolk, Southdown or Dorset Down, is frequently undertaken. This type of ram, used upon a half-bred ewe, again invokes the benefits of heterosis, assisted by the relatively high degree of homozygosity for carcass characters that has been attained within the Downs breeds most commonly used.

Cross-Breeding in Dairy Cattle

It is the crossing of the comparatively low-producing dairy zebu breeds with their natural heat and tick resistance, with the high-producing European dairy breeds that has proved most useful. One of the earliest of the long-term selections was undertaken in the West

Indies with the crossing of the Jersey with the Sahiwal. After many years of selection, this became known as the 'Hope-Jamaican'.

More recent attempts to create high-milk-producing cows for the tropics have been undertaken in Australia, where the Jersey and the Friesian have been crossed with the Sahiwal. From the selected half-breds, two new breeds or types have been established. One is called the 'Australian Milking Zebu', produced by workers in New South Wales over a long period of selection for heat and tick tolerance. The second is the 'Australian Friesian Sahiwal' selected in tropical Queensland and already yielding 8.8 per cent as much butterfat as the pure Friesian and showing far greater tick resistance.

Large numbers of the Australian Milking Zebu have recently been sent to Malaysia to start a new dairy industry.

There have also been extensive experiments to see the effect of crossing various European breeds of dairy cattle. Carried out mainly at Government-financed institutions in the United States, one of the most extensively reported was begun in 1949 at the University of Illinois. Here the Holstein was crossed with the Guernsey. Losses from disease were notably reduced in the crossbreds, 13.4 per cent for F_1 crossbreds compared with 32.7 per cent for the purebreds. None of the various traits associated with reproduction (calving interval, gestation length, interval from first service to conception, calving date to first service) showed any significant effect. The production records for first and second lactations showed positive heterosis effect for milk yield, butter, solids-not-fat and protein of 7.3, 6.4, 4.7 and 7.4, respectively; the crossbreds were also heavier. This led to the conclusion that under practical conditions the crossbred would be more profitable. However, in other American trails, in one of which Friesians, Channel Island breeds and Red Danish cattle were compared with their crosses, it was concluded that three-way crosses had only a slight advantage over the two-way cross, and that, in several instances, yields and milk quality were not significantly better than from the foundation breeds.

In temperate regions, the widespread use of improved methods of progeny testing bulls, associated with artificial insemination, has continued to raise both milk yield and quality of the common dairy breeds, of which the Holstein-Friesian is the outstanding example. If the cost of replacement heifers becomes very high and incidence of chronic disease, such as mastitis and infertility, much greater, the longer reproductive life-span and greater liveability of the crossbred may have to be resorted to.

Beef Production

Reference has already been made to the creation of new breeds of cattle adapted to heat. Much of this work has been undertaken in the southern and south-western United States. A survey of the results from trails made over a number of years in these areas shows that when the zebu beef breed known in America as the 'Brahman', was crossed with Hereford, Angus, Shorthorn or Charolais, there was marked superiority shown in the growth of calves up to weaning. Expressed on the basis of pounds weight, the crossbred calves were 15 per cent better than the average of the pure breeds. The F_1 crossbred cows performed better than the pure breeds that had been used in the cross. There was an average advantage of 14.3 per cent in the number of calves weaned, which, with the greater weight of calves at weaning, adds up to a considerable advantage for the policy of cross-breeding.

With the crossing of different beef breeds for use in temperate regions, the advantage is not so marked. American data show that, while just over 10 per cent more conceptions to first service were obtained with crossbred calves, the loss to time of calving was at least as high as in purebreds. One large survey showed that the average number of crossbred calves born was 5.2 per cent higher when purebred cows produced crossbred calves than when they gave birth to purebred calves. The crossbred calves gained weight at about a 5 per cent faster rate and made about 5 per cent more economical gains than purebred calves.

A three-way cross is sometimes used. The crossbred cow is mated with a bull of a third beef breed. There are not so many critical studies published as there are with the simple halfbred but, in some instances, there has been increased advantage, in the order of 12-14 per cent, expressed as pounds of calf weaned for each cow mated. When this is added to the 10-12 per cent advantage for the crossbred over the purebred calves, it could be said that the three-way cross has a 20-26 per cent superiority.

Most of the hybrid vigour seems to find expression in the calves before weaning because of their greater physical fitness and vigour. There is also a beneficial effect upon the carcass when a large breed, such as the Charolais or Chianina, is crossed with a smaller earlier maturing breed, for example, the Angus or Devon. Not only will the carcass be heavier, but there will also be a smaller proportion of fat at maturity.

Use of beef bulls with dairy cows

Beef bulls, such as the Hereford and Angus, are frequently crossed on to dairy cows, more particularly heifers, when conception is desirable at an early age. There is less danger or dystokia, or difficult birth, and a valuable calf is available for beef production. Crossbred heifers may be kept for use as dams in single suckler beef herds because of their greater capacity to provide plenty of milk for the calf. The full halfbred may not be sufficiently hardy to withstand harsh winter conditions in some northern countries.

Pig Production

The increasing pressure from the public for pork and bacon with a minimum of fat, has meant a search for breeds with the capacity to grow fast, make efficient use of high-energy feedingstuffs and reach marketable weight on a restricted diet before fat conversion becomes excessive.

It has meant the virtual disappearance of many breeds, such as the Berkshire, Large Black, Tamworth, Saddle-back, and concentration on the white-skinned breeds such as the Large White, Landrace and Welsh, with the Pietrain and Hampshire being extensively used in Europe and the U.S.A. Many commercial producers do not maintain pure breeds, preferring to make use of what hybrid vigour can be obtained by crossing. This has progressed to the extent that 85 per cent of pigs marketed in the U.S.A. are crossbreds. Many are content with a two-way cross. If a particular type of half-bred sow has been found to possess high fertility and good mothering ability for a particular environment, tested boars of a third breed can be used to enhance the degree of heterosis. With such a system, it is necessary periodically to obtain new gilts and boars.

By using a 'criss-crossing' system, it is possible to retain much of the beneficial effect of hybrid vigour for a number of generations without having to obtain fresh breeding stock. Suitable gilts or sows and tested boars of the two breeds are obtained from reputable studs. Following the first mating, the required number of gilts are selected and mated to a boar of one of the original breeds. From litters of this mating, the selected gilts are mated to a boar of the other breed.

A practical way of obtaining the advantage of the additional degree of hybrid vigour available from a three-way cross, is to use attested purebred boars of each of the breeds decided upon in rotation.

It has been clearly shown that the main advantages in cross-breeding pigs are in the larger litters, reduced mortality to weaning, with faster growth, particularly up to weaning.

Poultry Production

Economic forces have always played a dominant role in this industry, whether in the breeding, production and marketing of broilers, eggs, turkeys or ducks. The margin of profit upon any single unit is small, so that both commercial producer enterprises and breeding companies providing rearing stock operate on a large scale.

The breeding programmes are, therefore, often complex and large scale, making use of various combinations of inbred lines, which themselves are likely to have stemmed from particular breeds. In the process of creating a new hybrid as broiler or layer there will be a number of trail crosses before a final decision is made upon the combination of lines.

The large number of discarded males or females in the development of the qualities required in the two sex lines, many of which may be unmarketable, is one of the reasons why costs of finding a better hybrid are high. In the end, it may be that a four-way cross is decided upon, one for the production of the male line, the other for the female line.

If the programme is designed to produce a new strain of layer, it is likely that fertile eggs or day-old chicks will be sold to the individual producer. If it is a broiler line, the male and female parents may be sold to a large producer who holds a franchise from the breeding company that permits him to produce the final hybrid chick. But the grand-parent stock, the vital four lines in which reside the special combination of virtues, will be closely safeguarded and not sold.

General

The move away from the use and, therefore, the continued breeding of a multiplicity of pure breeds of horses, cattle, sheep, pigs and poultry continues. With the more rigorous selection of characters of economic advantage rather than of appearance, there is likely to be demand for only those few breeds found to give the best basis for cross-breeding.

That is the situation to-day. The exact nature of the traits that will be desirable in livestock in the future is unknown. This is a cogent reason for preserving whatever breeds remain in small but viable numbers. They represent gene combinations that may one day be needed.

Numbers and trade names have long since replaced the name of a particular breed when one wishes to identify a type of broiler or laying hen. Cattle and horses, even sheep and goats, with their comparatively long generation interval and meagre reproductive ability are not so genetically malleable. Progress towards greater productivity is notably slower. The developing non-surgical technique of embryo transfer, associated with more precise control of the reproductive cycles in the larger animals will certainly speed the propagation of superior lines.

5

DEVELOPMENT

MOVING TOWARDS INTENSIVISM

The growth and development of an animal shows many characteristics of its species, one of the most important being the fact that different periods of time are taken to reach marketable weight. For example, a broiler today is sent for slaughter at 8 weeks, having attained a weight of 2 kg. In the most intensive form of beef production a suitable animal can be marketed by 12 months of age. It means keeping the rapidly growing calf indoors all the time and feeding it concentrated high-energy food such as is to be found in a diet consisting of approximately 90 per cent cereals. Extra beef from a heavier animal can be produced more economically by the utilization of pasture during the spring and summer months, but it will take 15-18 months to attain an acceptable market weight.

It has been one of the principal aims of breeders to shorten the time taken by most species of domestic and farm stock to reach maturity. Geneticists found that the rate of weight gain was a relatively highly inherited characteristic. In any region where there are large numbers of a particular breed of livestock and adequate facilities for measuring and recording growth, there has been rapid progress towards reducing the age at which the majority of that breed will reach maturity. This has happened to Friesian cattle and Large White pigs in the U.K., with Merino sheep in Australia and Hereford cattle in the U.S.A.

Development is a continuous process from the embryo to the age of maturity when growth normally stops. Throughout the whole process, the parts of the body, composed of varying proportions of the main tissues, bone, muscle and fat grow at different rates. It is the brain and the nervous system, along with the digestive tract, that develop first in the embryo and, by the time the animal is born, are relatively well developed, compared with the muscles and bones. After birth, it is the muscles that grow fastest, followed by bone tissue, and only much later is fat laid down in any quantity. The important question of which factors affect carcass quality will be more fully examined later in this chapter.

Dressing Percentages

The end product of much animal production is saleable meat. It is important to understand which factors are responsible for the determination of the proportions of lean meat (muscle), fat and bone. There is, of course, a percentage of the animal's liveweight at slaughter that is not saleable as part of the carcass, although having commercial value for other purposes. This includes the skin, hair, wool, blood, caul or mesenteric fat, certain glands, meat and bone meal. The dressing-out percentage varies between species and depends upon such factors as age, sex, nutritional level and ante-mortem treatment. A younger animal will kill out at a higher percentage than older animals. The degree of fatness is the biggest variable in mature animals, as the excessively fat animal will dress out at a lower percentage due to the greater loss in mesenteric or caul fat. Whether or not the kidneys and their surrounding fat ('knob fat') are left in the carcass, can also make a difference.

The manner in which ruminants are weighed before slaughter can notably affect the dressing percentage, particularly through the large variation in gut fill. To obtain a weighing that can be regarded as fairly representing body weight, it is necessary to withhold all feed for 24 hours but to give access to water. An alternative method, although seldom practically possible, is to weigh on number of occasions on successive days and take the mean. The way that diet can affect dressing percentage was shown by a comparison made by Butterfield (1977) when he fed calves solely upon various levels of milk and compared them with similar calves fed upon a diet of roughage as well as milk. The latter showed dressing percentages falling from 55 per cent at 4 weeks to 46.1 per cent at 22 weeks, whereas the calves given only milk increased in dressing percentage throughout the experiment. Brahman bulls kept on a high forage diet had carcasses that dressed out at 3.8 per cent less than bulls of the same weight given concentrates. Differences between bulls, steers and heifers killed at comparable weights and fed similarly can depend upon methods of dressing the carcass. If the kidney fat is removed, bulls have been shown to be superior to their castrated twin, but in other studies, when kidney fat has not been removed, dressing percentage has been similar.

Different breeds will show variation in dressing percentage. It can be said that the traditional British beef breeds, Shorthorn, Angus, Hereford, will dress out at a higher figure than dairy breeds because of the greater degree of subcutaneous fat. There is a good deal of evidence that the Charolais has a higher dressing percentage than several other breeds, one reason apparently being its lighter hide. Because of the many variables, Berg and Butterfield (1976) conclude that 'with present knowledge we must conclude that carcass weight, live weight and dressing percentage are inadequate descriptions of value in a beef animal'.

Sexual Development

An important stage in each animal's progress towards maturity upon which management can have an important effect is sexual development and reproductive processes.

Puberty and mating

The development of the organs of reproduction reach a stage of functional development some appreciable time before the farm and domestic animals reach the age of maturity.

It is from this time onwards that the male will be producing mature spermatozoa and the female will have periods of oestrus and an egg capable of being fertilized will be released. The female may of course be mated at an appropriate stage during one of them. The right time to do this depends upon knowing at what stage during oestrus ovulation occurs.

The manner in which an animal is fed has a considerable effect upon the age at which an animal reaches puberty. Ill-fed females, even if they belong to an early maturing breed, will be older before their sexual cycles begin. This can be of much practical interest in dairy cattle, for it is generally desirable to get a heifer into calf by as close to 15 months old as possible. In the more slowly maturing zebu dairy breeds and where climate and feed supply may interfere with the aim of providing optimal nutrition, age of first conception will be later. Very high energy and adequate protein inputs will permit fertile oestrus in European breeds of dairy cattle at about one year old. A higher incidence of difficult calving and a poor first lactation sometimes follow so that such precocity is not advocated.

Effects of oestrus

There are notable variations in the regularity with which different species of animals have oestrous cycles. The cow and the sow continue to cycle throughout the year at a length of approximately 21 days from one heat period to the next. The mare goes into a state of anoestrus during which cycles do not occur from the late autumn throughout the winter. With the lengthening of the daylight hours in the spring, oestrous cycles will recommence. In the ewe, it is the shortening of the days at the beginning of autumn that starts her cycling. While this is the pattern in a temperate climate, in tropical and subtropical countries, the comparatively small difference in the length of daylight throughout the year results in a blurring of such rhythms, so that all species can be bred throughout the year, but with increased fertility shown during the equivalent of spring and autumn in such parts.

Normal behaviour is interrupted during heat periods, including, with some individuals, reduced feeding. There will be some slackening of the rate of growth in young animals and a drop in production of cows in milk. If mating or artificial insemination is successfully accomplished, there will be no measurable increased demand for additional energy or protein during the early part of pregnancy. The heifer, the ewe lamb, and the gilt will continue their growth.

Pregnancy

The manner in which hormones control both the repeated cycles of the female waiting to be mated and the changes that occur with pregnancy, are complex. Some hormones are produced from the pituitary in response to signals from that nearby part of the brain known as the hypothalamus, and from the ovary, while others come from the developing placenta, where there is the vital connexion between the circulations of dam and offspring. Research during the past few years has added new hormones and identified many 'releasing factors' produced in the anterior pituitary.

The importance of exercising continuous careful control over the pregnant female's nutritional requirements is clear. She has to provide for her own continuing growth if it is a first pregnancy, or, as with a lactating cow, for continuing milk production, as well as for a developing foetus. Subsequently, she may have to suckle her offspring, or to embark upon a full lactation in the case of a dairy heifer. As the end of pregnancy approaches, the rate of growth of the foetus greatly accelerates. Not only is the foetus increasing in size and, therefore,

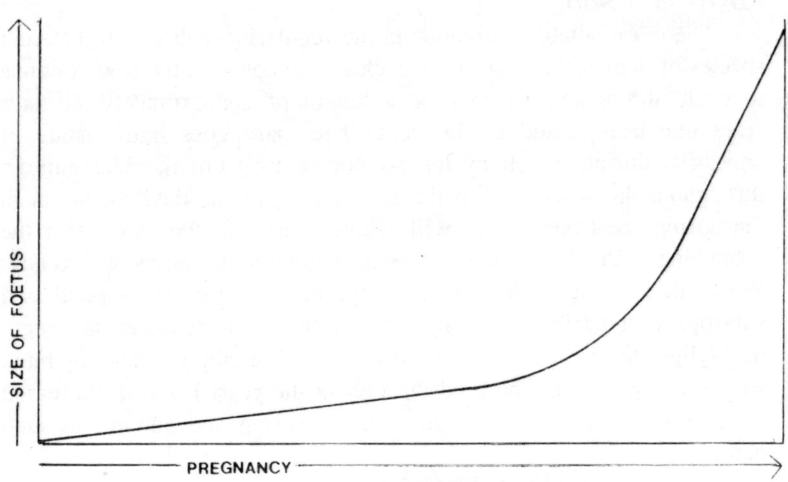

Fig. 5.1. Growth of foetus during pregnancy.

making demands for additional nutrients as the end of pregnancy approaches, but also the mammary gland is developing in preparation for lactation.

With the older dairy cow, who has to face the heaviest and most prolonged lactational demand, it has been shown that a steadily increasing level of feeding should begin towards the end of the preceding lactation and continue during the two months of the dry period usually allowed before calving. If the period of weight renewal and preparation for the next lactation is extended back to the latter part of the preceding lactation to make it a continuous process of energy supplementation through the dry period and calving up to the peak of the succeeding lactation, there is the best possible chance for the birth of a vigorous well-grown calf, a high lactation peak and no illness caused by inadequate available energy or mineral imbalance.

Care at parturition

Parturition or the birth of the young animal is clearly a critical time and special care needs to be taken. A record of the date of mating should be kept, but there are various physical signs shown by the majority of animals that indicate the end of pregnancy. Apart from the increase in size of the abdomen and the mammary gland, the vulva will shown some swelling, often accompanied by reddening. The animal tends to detach herself from the rest of the herd or flock to seek a place where her offspring can be born. It is wise practice at this stage with the mare, the cow and the sow to have them in separate accommodation where they can be comfortably bedded and be kept under quiet surveillance.

Most ewes lamb outdoors but in northern European countries they are increasingly being brought indoors for lambing, if not for the whole winter. The improved environment for both ewes and shepherd and the greater care resulting is responsible for reduced perinatal deaths.

Does are treated similarly in temperate countries where the improved milking breeds are used. In many subtropical and tropical places, kidding takes place wherever the flock is grazing at the time.

CATTLE

Birth of the Calf

Difficulties at birth, or dystokia, can occur with all species but is most common in the cow, particularly with the heifer having its first calf. There is a widespread tendency to plan for the first calf to be born when the heifer is only two years old and growth has by no means been completed. It is wise practice for heifers to be mated to a bull of a small breed such as the Aberdeen Angus or Welsh Black to try to be sure that there is no difficulty at calving.

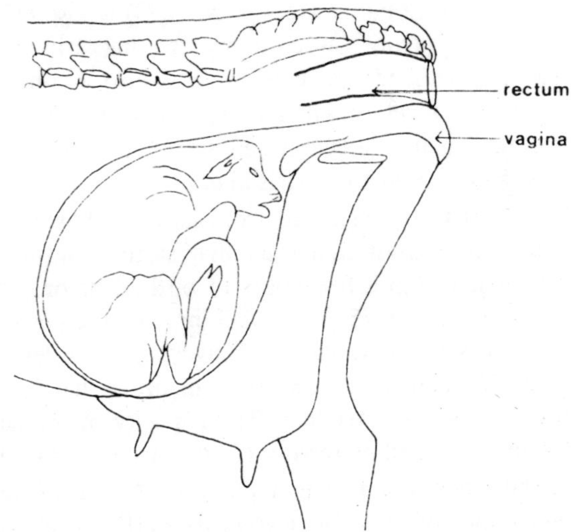

Fig. 5.2. Position of calf at term.

Dystokia

The commonest causes of difficulty at birth are the misplacement of one or other of the forelimbs of the foetus or the turning back of the head. Occasionally, the whole body may be reversed and the hind legs instead of the forelegs presented in the vagina. It is obvious that the first thing that must be done if the attendant sees that the foetus is in the wrong position is very gently, with a properly cleaned and lubricated hand, to press back the foetus and try to bring the bent limb forwards so that both forelimbs or both hind limbs are presented in the birth canal. It is important when traction is to be used to assist birth that this be not too great and pulling should synchronize with the cow's straining. If the amniotic sac or water bag does not appear after some hours or if there is any sign of the afterbirth appearing but the calf has not been born,. or if there is any discharge of putrid matter, a veterinary surgeon must be called at once.

Neonatal period

The calf will start breathing as soon as it is born and is free of the birth membranes and all that is necessary is for the attendant to see that the calf gets on to its feet and the mother licks it within a short time of birth. It is of the utmost importance that the calf suckles its dam within the first four hours of being born. The fluid produced in the udder, known as *colostrum*, contains valuable protective substances, immunoglobulins (IgA, IgM, IgG) whose effect is to give the calf immunity from infection by the bacteria that inevitably contaminate its surroundings. Colostrum is also a concentrated food, containing a high level of vitamin A, which also has the effect of helping the calf against infection and assisting its growth; it has high levels of fat and protein for nourishment.

If the calf is to be separated from its dam shortly after birth, or the cow dies, it is most important that stored colostrum should be given to the calf within a few hours of birth. It is only necessary for the calf to receive 1-1.75 litres (2-3 pints) of colostrum to obtain protection from subsequent bacterial invasion. It is desirable to leave the calf with its dam for 24-28 hours in order that it may have the full benefit of colostrum feeding. The capacity of the calf's intestine to absorb immunoglobulins rapidly diminishes after 24 hours.

Since the modern dairy cow has been selected for her capacity to produce quantities of milk far beyond the calf's needs, it is normally removed from its dam after receiving sufficient colostrum. It is then fed whole milk for a week or 10 days before being transferred on to a cheaper but nutritionally equivalent diet of milk replacer, and also given concentrates and hay. Growth rates on this regime, assuming freedom from disease, are similar to those attained by calves given whole milk. In some situations, such as may occur in developing countries, the cost of importing milk replacers may make it as cheap to feed whole milk to the young calf. It is a traditional practice in tropical countries to give calves whole milk for at least two months. Weaning may not take place until three month's of age.

Sheep

Birth of the Lamb

The ewe is a seasonal breeder with a pregnancy lasting approximately five months, and since she comes into season as daylight shortens, lambs are normally born in the spring. In many countries, the weather at this time is uncertain. This not infrequently has the effect, unless precautions are taken, of interfering with the intake of supplementary rations and making surveillance of lambing difficult. Losses amongst lambs can be high (10-30 per cent). Death may be due to chilling, which prevents the lamb getting on to its feed and suckling, or it may be too small and feeble to do so. Some ewes move away, instead of attending to her offspring by licking and nudging it on to its feet. A steady increase in the practice of bringing ewes indoors, or at least into adequate shelter for lambing, will lead to a reduction in losses at lambing.

Except under the still rare conditions of the most intensive form of producing fat lambs, when removal of the lamb from the ewe occurs after 24 hours, the common system is for the ewes and lambs to remain together during the two to three months that represent the length of lactation of most breeds of ewe. The growth rate and, to a large extent, the shape of the lamb, will depend upon its breed, whether it is one that has been selected mainly for wool production or for its meat qualities. In Spain and Portugal, there are still a considerable proportion of the ewes that are milked, mostly in order to be able to make special cheeses.

Pigs

The sow is only rivalled by the female rabbit, the doe, in her capacity of being able to produce a large litter of 8-14 piglets following a short gestation (114 days). It results in the piglets being born in a relatively undeveloped state, having little hair and being unable to control body temperature for the first few days. Thus in cool climates there will be the need to provide a source of additional heat. The adequately fed sow of a modern breed will regularly produce between 6 and 12 piglets and, in a large litter, there will often be one or two very small ones ('runts') with a poor chance of survival. Since the sow has only 12 teats, any piglets beyond this number are unwanted. There is a 'teat-order' of suckling established within 24 hours of birth.

The well•fed sow, with six pairs of teats, should be able to provide sufficient colostrum and then milk for all her litter. Good stockmanship demands that any newly born piglet that has any difficulty in finding a teat, is helped and then observed to be sure that it is successfully suckling. To reduce if not entirely to prevent losses from piglets being lain upon by the sow, her movements are commonly restrained by the use of a farrowing crate; at the same time, the piglets are encouraged to move a little away into a creep area which the sow cannot reach. Here warmth and special supplementary food is provided.

The young pig, once it is receiving enough milk from the sow supplemented by creep feed, undergoes a similar pattern of growth to all other farm animals.

Rabbits

This is the most fecund of the domestic species. The doe will begin breeding at about six months old and fertile mating will normally occur when she is put in with the buck. Ovulation depends upon copulation. The pregnancy only lasts 31 days and litters can vary widely, from as few as two or three up to 15 or 16, but with modern meat hybrid strains it is from six to nine. It is thus possible for a doe to produce 60 or more offspring during her first breeding year.

It is not surprising, in view of this short gestation, that the newly born kits are relatively undeveloped. They are hairless, slow-moving, deaf and have closed eyes. They are in need of the warmth and close confinement that is provided by the nest constructed by the doe just before kindling. In the small box provided in her cage, she will construct this nest from fur pulled out of the abdominal region.

The young rabbits remain in the nest for 7-10 days, but growth is rapid, notably when litters are of moderate size. Deaths amongst the smaller members of large litters are high from chilling and inadequate suckling. It is common commercial practice to remove the doe when the litter is about four week of age.

If the young rabbits are kept free of disease and adequate amounts of a suitable diet provided, they will attain a marketable weight of 2-2.5 kg within 8 weeks. As in the marketing of other forms of meat, it is the uniformity of carcass, both in size and quality, that determines the level and stability of the price paid. While the efficiency of food conversion of some recently produced strains of meat rabbits rival those attained by the broiler chicken, there is room for more intense selective breeding for rate of growth and efficient food use. The demand of the fur trade is for white pelts.

General Growth Pattern

Rates of growth, as measured by weight increase, vary not only with the species of farm animal and with the breed, but with strains

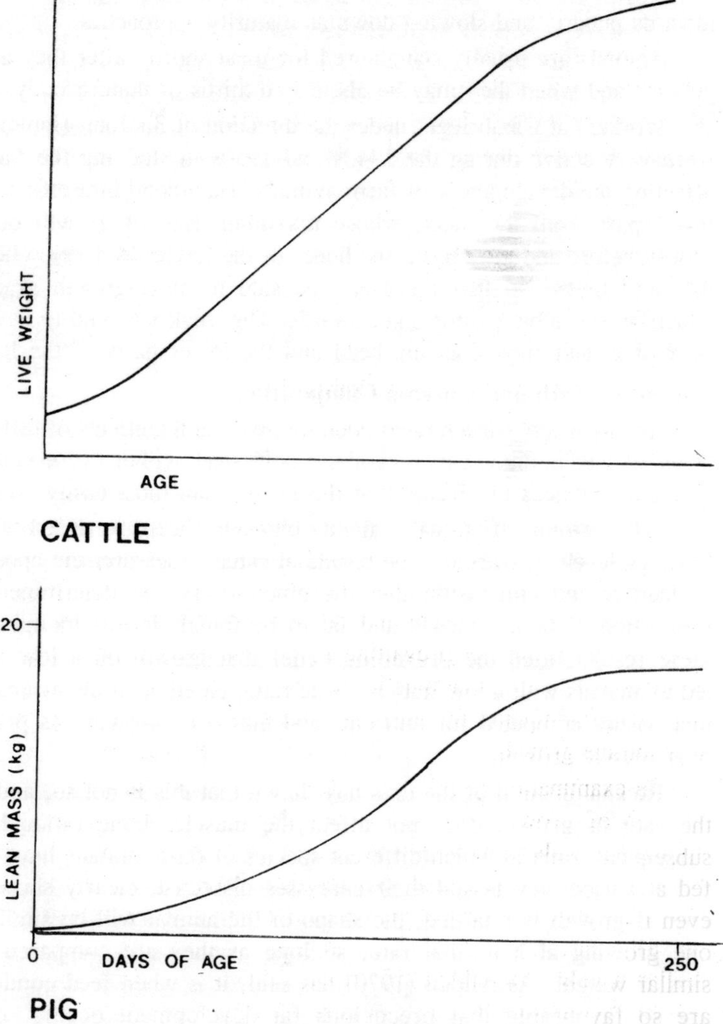

Fig. 5.3. Typical growth curves for cattle and pigs.

or lines selected within a breed. A dairy heifer will not increase in weight as fast as a beef bull calf, and there are strains within such breeds of pig as the Large White or Landrace that make faster growth than others.

It is also a characteristic of growth in farm animals that, so long as the nutrition is at an optimal level, the rate of gain is constant over a comparatively wide weight range. In the pig, this is from 20-120 kg and in beef cattle from 100-480 kg, with a burst of acceleration towards puberty and slowing down as maturity approaches.

Animals are usually slaughtered for meat shortly after they attain puberty and when they may be about two-thirds of mature body size.

Workers at Cambridge, under the direction of Sir John Hammond, were very active during the 1940s and 1950s in studying the factors affecting the development of farm animals. Hammond himself showed that, apart from the head, whose maximum rate of growth occurs shortly before and after birth, the bones of the forelimbs develop before the hind limbs, so that there can be said to be a growth gradient which proceeds backwards and upwards. The trunk was said to develop fully at a later time than the head and the lower parts of the limbs.

Rate of Growth and Carcass Composition

Attention was concentrated upon discovering the effects of different levels of feeding upon carcass composition, since within various classes of animals this is the factor that the farmer can most easily control.

When groups of animals, mainly pigs and sheep, were fed widely differing levels of nutrition, the results of carcass measurement appeared to lead to the conclusion that the amount of food determined the proportion of bone, muscle and fat to be found. It was thought that these results fitted the prevailing belief that growth on a low plane led to animals with a low muscle: bone ratio, based upon the assumption that tissues competed for nutrients and that bone growth has priority over muscle growth.

Re-examination of the data has shown that this is not so, and that the rate of growth does not affect the muscle: bone ratio. Many subsequent trails in which different species of farm animals have been fed at various levels and their carcasses dissected, clearly show that even if growth is retarded, the shape of the animal will be similar to one growing at a normal rate, so long as they are compared at a similar weight. As Allden (1970) has said, it is when feed conditions are so favourable that precocious fat development occurs, or so restricted that growth of bone continues at the expense of other tissues, that the shape of an animal may be altered, though usually to a minor degree.

The allometric equation $y = bx^a$, where y = size of organ or part, x = size of the rest of the body, a = growth coefficient of the organ or part, has been widely applied. When used in beef cattle to compare the growth of muscle, fat and bone, it was found that over the entire growing period, the growth coefficient for bone is less than unity, for muscle it is just above unity, but for fat is high, from 1.5 to 2.0

Bone develops early in life, so that at birth the ratio of muscle to bone may be 2:1. But muscle subsequently develops at a faster rate than bone, with the latter continuing at a slow, steady rate. There is not much fat in an animal at birth but it steadily increases, most

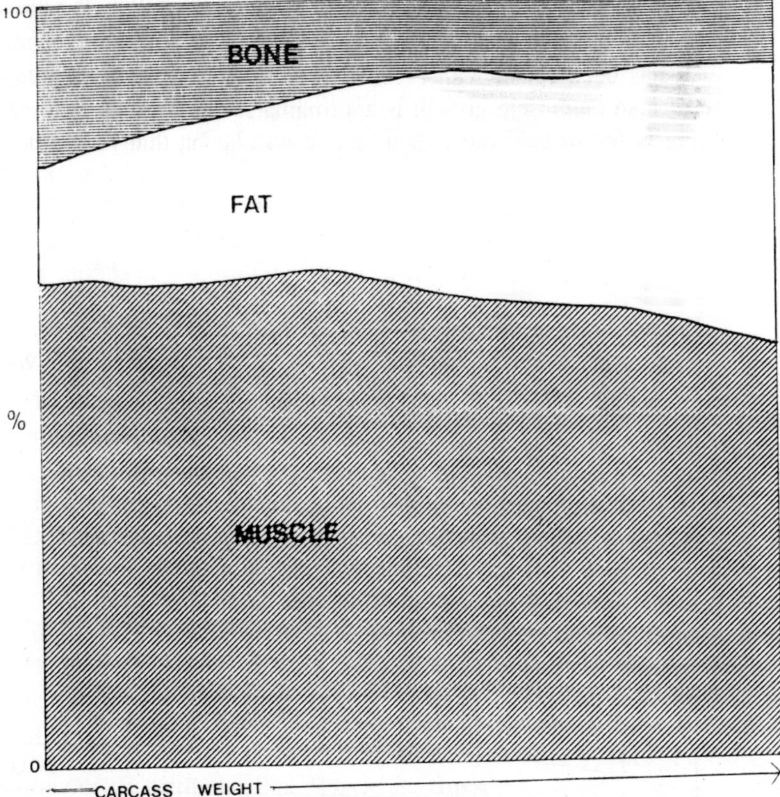

Fig. 5.4. Proportions of tissue as percentage of carcass composition.

markedly in the pig. Even in sheep and cattle, the rate of fat deposition can exceed the growth of muscle in the mature animal under a feeding regime of high energy intake.

Growth in Pigs

Pigs differ from ruminants in the way food is utilized. From the time the young pig is weaned, any excess food energy is laid down as fat. It is for this reason, since excess fat is not wanted in the carcass, that rations for modern commercial strains of pork or bacon pigs are restricted. The accummulation of fat continues at an ever faster rate as the pig increases in size, according for 10 per cent in the carcass of the young pig of 10 kg weight, but for 30 per cent of the body of the 100-kg fattener, even if kept on a restricted diet.

In contrast, protein makes up about 15 per cent of the body weight, a little more in the young pig and slightly less in the older animal.

It has been found with present types of pigs, that the maximum rate of lean or muscle growth is approximately 450 g daily, so that if the pig is fed to gain more than that, it will be fat that is laid down.

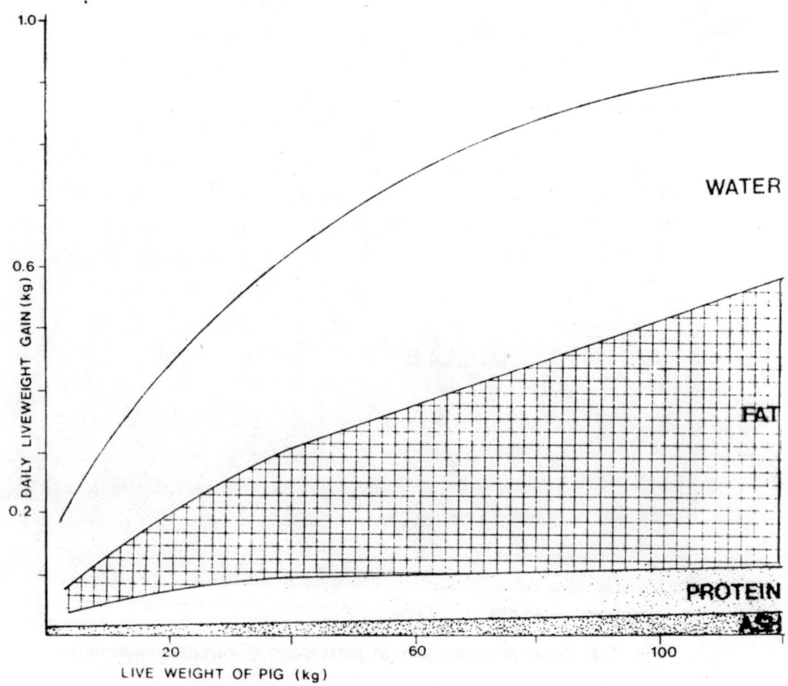

Fig. 5.5. Changes in proportions of water, fat, protein and ash with growth in pigs.

It is stated that the composition of tissues below 550 g daily gain will be approximately 80 per cent lean: 20 per cent fat.

The growth of muscle, i.e. mainly protein, involves a large proportion of water. Mammalian muscle protein contains 75-80 per cent water, whereas fat has only about 10 per cent water. The proportion of water in the muscles of the young animal is at a high level, decreasing in the mature animal. There is also variation in the ratio of protein to fat in each unit of weight as the pig ages. In the small 10-kg pig it is 1:1 but by the time a baconer is ready for slaughter, say at a weight of 100 kg, it has widened to 1:4.

The efficiency with which food is converted to tissue growth falls in all animals as they get older and fatter. In the pig, food conversion efficiency (FCE) measured as the amount of food needed to produce unit gain in weight, changes from 2:1 in the 20-kg pig to 4:1 in the 100-kg pig.

Following slaughter, there are a number of parts of the pig that have entered into its weight measurement during its growth that are discarded, as we have seen. There has been increasing unwillingness amongst most people to accept more than a small amount of fat in either pork or bacon.

It is now the routine procedure in the abattoirs of many countries to base payment for the carcass upon a grading system that not only takes note of its weight and certain linear measurements, but also ascertains the thickness of the backfat. This is believed to be a reliable indication of the quantity of fat in the whole carcass. An introscope is inserted over the mid-point of the eye muscle and the reading, called P_2, becomes part of the grade assessment.

There may also be down-grading because of off-flavours in the carcass caused by the careless inclusion of a contaminated ingredient in the ration, such as badly prepared or rancid fish-meal. Of recent years there has been an increase in the number of carcasses, particularly from the Pietrain breed, which are especially prone to react to stress, that show musculature that is pale, soft and exuding fluid (p.s.e. meat). This appears to have been a side effect of intense selection pressure for rapid growth.

Growth in Cattle

A series of studies on the growth of beef cattle during the past few years have greatly extended our knowledge of the various factors responsible. They have confirmed the importance of considering groups of muscles, their growth and total contribution to the lean mass available from the carcass, free from fat. When this is done, it is immediately clear that there is a remarkable degree of uniformity of muscle weight distribution in cattle of all breeds and types.

The classification of groups of muscles shows that those thought to be gastronomically the most delectable and therefore commanding the higher prices wherever meat is sold, are not particularly well developed in cattle. These are the muscles around the backbone and of the hindquarters.

Berg and Butterfield and their co-workers (1970) have shown that it is in such animals as antelopes and the moose, which are agile animals, that there is the largest relative development of distal limb muscles; while in the small domesticated animals, the sheep and

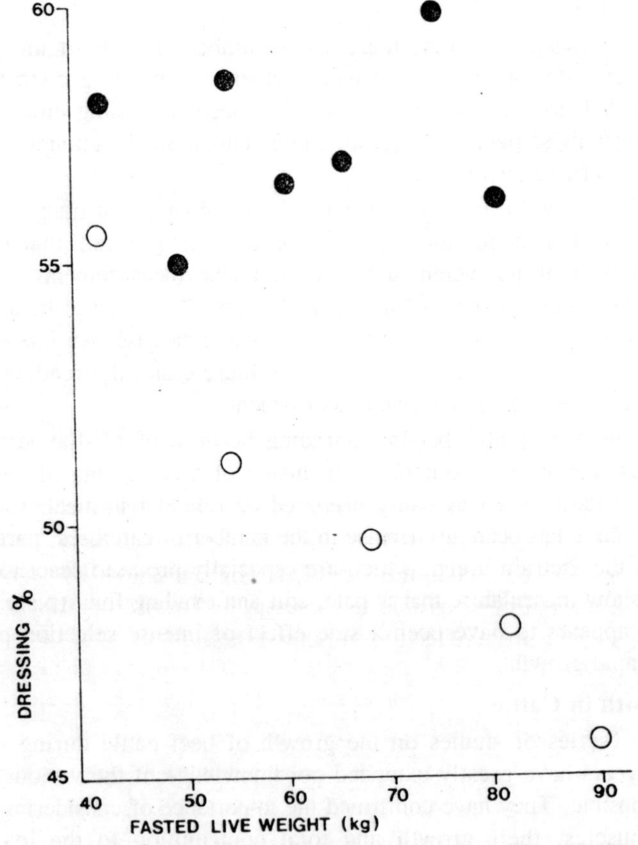

Fig. 5.6. Effect of diet on dressing percentage with calves fed on (●) milk only or (○) milk + roughage.

particularly in the pig, the expensive loin muscles are much better developed.

It is suggested that the reason for the greater development of the abdominal muscles in cattle lies in the fact that, in the long period since domestication, they have been used as grazing animals for the consumption of large amounts of fibrous pasture plants. The effect that diet can have on dressing percentage due to the much smaller weight of the digestive tract of calves fed milk only.

Factors Controlling Carcass Composition

What ultimately determines the value of a carcass is the quantity of edible cuts of meat that can be sold, associated with which has to be a consideration of the proportion of those of greatest value. Those that eat meat of whatever species would agree that the superior carcass has a maximum amount of muscle, a minimum of bone and an optimum amount of fat, for consumer preference varies over the latter.

There are three factors that influence the growth of these tissues: genetic, sex and nutrition.

Genetic influences

These are to be seen most clearly in the long-established breeds, but can also be revealed in certain strains or types within breeds and in crossbreds.

Some breeds have been selected for early maturity. This means that their maximum weight tends to be less than in late maturing breeds and that the increase in the rate of laying-down of fat relative to muscle begins at an earlier age.

An examination of data provided by the Royal Smithfield Club showed that Hereford steers fattened at an earlier age than Friesians in relation to weight.

When the contribution made by muscle and bone was separately considered, it became clear that the Herefords also had more muscle relative to bone than the Friesians. There are good reasons to believe that it would be possible to increase the amount of muscle relative to bone by genetic selection in other breeds.

Sex

It is a common observation that at the same weight and age, heifers will fatten to a greater extent than steers. It is equally clear that with good feeding, bulls grow faster than steers and steers faster than heifers.

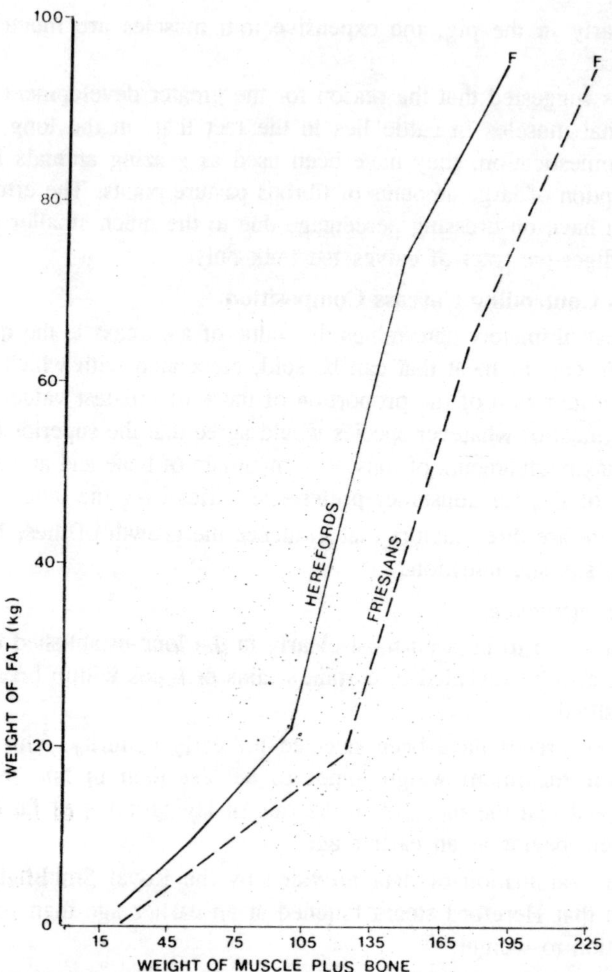

Fig. 5.7. Laying-down of fat relative to muscle + bone in Herefords and Friesians.

By using monozygotic twins, it was possible to show that castration slowed liveweight increase by 10 per cent and muscle growth by 17 per cent. One of Berg's studies with a large number of animals kept under comparable conditions, showed that, expressed as weight gain per day, bulls gained 1.070 kg, steers 984 g and heifers 869 g. That this was mainly muscle and bone rather than fat became clear when it was seen that bull carcasses yield 482 g per day of muscle and bone compared to 390 g and 326 g for steers and heifers, respectively. When muscle was plotted against bone, there is little difference between

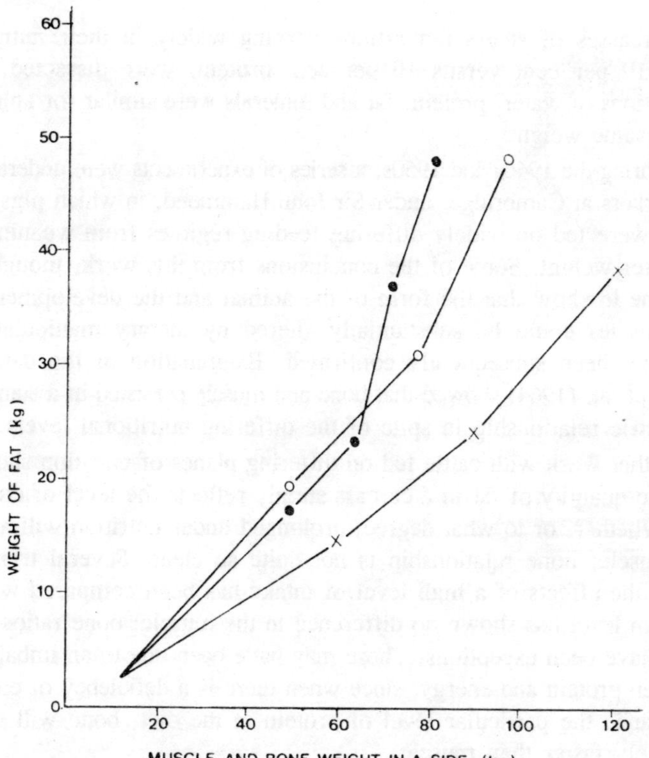

Fig. 5.8. Laying-down of fat relative to muscle + bone in (●) heifers, (○) steers or (×) bulls.

the sexes when they are compared at similar stages of development of muscle and bone.

It appears, then, that the main differences between the sexes is the fact that the impetus for fattening comes before that for muscle growth, thus occurring at lighter weights in heifers than in bulls, with steers intermediate. The fact that bulls have more muscle relative to bone is probably due to the prolongation of this impetus before the final fat deposits occur.

Nutrition

The general effect upon carcass composition from known levels of feeding has been studied since last century. Apart from the pioneer work of Lawes and Gilbert at Rothanstead in England (1859) work on a larger scale was done by Jordan at Maine beginning in 1893 and by Haecker from 1908 at Minnesota. They were able to show that when the carcasses of steers fed rations varying widely in their nutritive ratio (19 per cent versus 10 per cent protein) were dissected, the proportions of water, protein, fat and minerals were similar for animals of the same weight.

During the 1940s and 1950s, a series of experiments were undertaken by workers at Cambridge, under Sir John Hammond, in which pigs and sheep were fed on widely differing feeding regimes from weaning to slaughter weight. Some of the conclusions from this work, thought at the time to show that the form of the animal and the development of the muscles could be substantially altered by dietary manipulation, have not been subsequently confirmed. Examination of the data by Elsley et. al. (1964) showed that bone and muscle persisted in a standard allometric relationship in spite of the differing nutritional levels.

Other work with cattle fed on differing planes of nutrition showed that the quantity of fat in a carcass simply reflects the level of energy fed. Whether, or to what degree, prolonged under-nutrition will affect the muscle: bone relationship is not quite so clear. Several trials in which the effects of a high level of intake has been compared with a medium level has shown no difference in the muscle: bone ratios, but there have been exceptions. These may have been due to an imbalance between protein and energy, since when there is a deficiency of energy to balance the particular level of protein in the diet, bone will grow relatively faster than muscle.

The effect of a period of under-nutrition, followed by adequate levels of feeding, has been studied. When there was an arrest of growth at a young age, followed by normal growth, and the animals killed at equal slaughter weights fixed on an ascending scale, there was no difference in carcass composition. The latter was simply related to the size reached. These results came from a trial in which none of the animals had reached the age when fat deposition occurs.

The feeding of high energy diets in large amounts to farm livestock will result in greater fat deposition in relation to muscle and bone, for this is the traditional way to achieve the smooth 'finish' to steers that has hitherto been demanded by the butchers of many countries, or to win any class at a show.

If weight loss is imposed on fattening steers, there ensues rapid loss of fat, and if low intake continues there will be about equal weight loss from muscle and fat until the fat reserves are exhausted. Further weight loss must increasingly involve muscle and, to some extent, bone.

Realimentation will restore muscle and bone to normal levels first, but the re-establishment of full fat percentage takes a long time.

An illustration of the practical effects of unsupplemented grazing on rate of growth of steers, their market value and carcass composition has been presented by Morgan (1979). Only by giving supplements of hay (16.8 per cent crude protein, digestible organic matter 65 per cent in the first year, and 12 per cent crude protein, digestible organic matter 49.0 per cent in the second year) along with oats, so that their growth could continue at 0.8 kg per day during the autumn and winter, could they be ready for market at 19 months old, when their mean live weights were 420 and 450 kg in years 1 and 2, respectively. It took the unsupplemented group until the age of 27 months to reach the same average liveweight.

It was found that the supplemented group had a greater depth of subcutaneous fat and the boneless sides contained more chemical fat. The group 1 carcasses were shorter, as were the radii and ulnae, and in one year the Longissimus dorsi muscles (fillet steak) were much more tender. This was not repeated in the second year. The grading discriminated against the carcasses with long bone dimensions and less than 6-9 mm fat depth over the loin. Yet the group 1 carcasses had the lowest proportion of lean meat in year 1, and the lowest proportion of muscle in year 2.

Other recent work has confirmed that reducing the feed intake of steers reduces carcass fatness at the same carcass weight and grain feeding increases carcass fatness relative to grazing on pasture alone. It is also known that steers that have lost weight but have later recovered their original liveweight, have less fat in their carcasses than steers maintained at the same liveweight during the same time.

It is as well to remember that the influence of the concentration of the energy in food upon both rate of weight gain and fat content of the carcass has been the basis of the most intensive method of beef production, the so-called 'barley beef', because by allowing only about 500 g of forage daily, along with 90 per cent rolled barley, supplemented with protein in the earlier stages and minerals and retinol throughout, even the slower maturing breeds of cattle can be ready for slaughter by 12 months old.

Inadequate intake of protein in the young ruminant will also affect growth and carcass composition. In experiments with lambs of differing ages, it was shown that low levels of protein resulted in higher levels of fat in the carcasses; for example, at 10 and 12.5 per cent protein

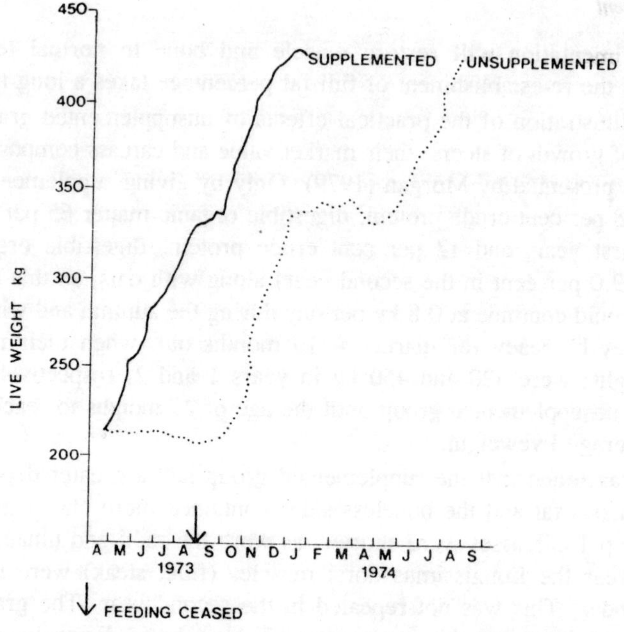

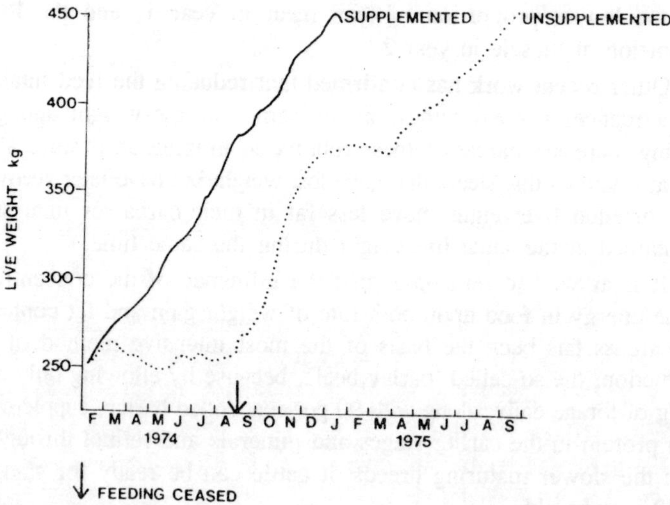

Fig. 5.9. Growth curves of supplemented (—) and unsupplemented (.....) groups of steers.

there was more fat in the carcasses of lambs of the same body weight than when 15 per cent protein was fed.

It is also interesting to see that the proportion of fibre in the diet can have at least a marginal effect upon the carcass, since it has been shown that, irrespective of the total intake of energy and protein, when some calves were fed whole milk and others a fibrous diet, the latter developed heavier abdominal muscles.

Muscle Development

It was at one time held that changes in the proportion of the muscles in a carcass was a reaction to changes in the relative proportions and angles of the bones. This has been shown not to be true. The forequarters of bulls contain more muscle weight than those of steers, because of the greater development of the neck muscles in particular. But the percentage of bone in the forequarter carcasses of bulls and of their castrated twins has been shown to be the same.

Muscles grow at different rates. Those that are most closely related to the skeleton and are smaller, such as the loin muscle, have a lower growth impetus than those that are superficial, as are the muscles of the leg.

Berg and Butterfield (1976) state that the available evidence supports a phasic variation in muscle growth rates.

Phase 1 is concerned with the ante-natal growth where the genetic template for the species ensures that at birth the animal can meet the environmental stresses that would occur in the wild state, muscles being stimulated by passive tension from skeletal elongation.

Phase 2 is the immediate post-natal period and is one of great change in muscle weight, continuing until there is doubling, and even quadrupling of the weight of some muscles. It is largely influenced by function and is thus similar for all farm animals, unless there is much difference in diet, when muscles of the abdomen will differ in relative growth.

Phase 3 is the pre-pubertal and adolescent period when the muscles grow at an almost uniform rate, giving a large increase in size with little change in relative weights since function remains similar.

Double-muscling in cattle

Studies with double-muscled or 'Culard' cattle show that there are changes in the relative size of muscles and that the growth of bone and muscle are essentially independent of each other. This inherited genetic trait is found most commonly in such heavy slow-maturing breeds as the Charolais and South Devon. When the weights of individual muscles of seven hypertrophied and seven normal animals at equal live weights were compared, there was 22 per cent more weight of muscle as a total. This was unevenly distributed, 25 per cent more in the muscles of the forelimb, 24 per cent more in the muscles of the hind limb and 16 per cent more in the muscles of the thorax and loin.

It has also been shown by French workers that double-muscled cattle have the greatest reduction of bone weight in the proximate part of the limbs, which is the area where there is the greatest increase of muscle weight compared to normal cattle, emphasizing the independence already mentioned.

It would be altogether too hasty to conclude that it is desirable to select for this trait. Inevitably there are very serious difficulties at parturition. The imposition of the pain, distress and economic loss makes the use of Culard animals for breeding an unacceptable risk.

Influence of Breed on Muscle Growth

In much the same manner that altering nutrition can apparently make obvious differences in the carcass, so it would seem that the different shape and rates of growth of a wide range of breeds of cattle would change carcass characteristics. Following a large study of 170 cattle that included representatives of dairy, beef and crossbred bulls, steers and heifers, it became clear that, in spite of their marked differences in appearance, there was an impressive uniformity of muscle weight distribution. There were only two muscle groups in which there were any breed differences. One was the muscles of the abdominal wall and the other the intrinsic muscles of the neck and thorax. The first of these have already been shown to be influenced by the nature of the diet; the second reflect the degree to which the muscles have progressed towards compositional maturity. It was the Jersey bulls that gave the greatest weight, presumably because they are so early maturing but at lighter weight. The Holstein steers although the heaviest, had lightest neck and thorax muscles suggesting that they were less compositionally advanced than other breeds.

The Butcher and the Carcass

When it comes to the carcass that the butcher has bought and cut up for his customers, he can be faced with a dilemma. He would like to have tender loin and proximal hind-limb muscles, with some interlobular fat within the main muscle groups, characteristic of a traditional beef breed such as the Angus or Hereford. But the cuts from these quickly grown early-maturing breeds require much trimming of surplus fat and the weight of lean meat will be less than in the larger slower-growing draught breeds, such as the Charolais. The high price of beef, relative to pork, poultry and, in most countries, mutton and lamb, associated with the increasing availability of small pre-packed lean joints to be bought in the supermarket, indicate a move away from traditional fat-enshrouded cuts.

There may be increased future use of the early maturing British breeds in crossing with the large draught breeds of Europe to combine the desirable qualities of being able to grow fast whenever energy/protein inputs can be high and to incorporate the greater muscle development of the larger breed. In sub-tropical and tropical countries where suitable housing when necessary and adequate feeding can be provided, British beef breeds crossed with zebu breeds have long proved of much value.

The increasing use of mixed breed crosses such as Colbred, Damline, Perendale, and many others to dams, crossed with rams of early-maturing mutton breeds or other selected crossbred rams for the production of fast-growing lean lambs is an illustration of the same trend. Similar breeding programmes are followed with pigs and broilers.

Comparative Studies of Muscle Growth

It is instructive to realise that in spite of the determination of so many breeders, the important characteristics of muscle size and distribution in cattle have not been altered. They have achieved obvious alterations to the external appearance and of fattening patterns, but that is all. Presumably this emphasizes the power of the forces of evolution exerted over so great a time-span.

In pigs

When the development of individual muscle groups of different breeds and sexes are compared, there is very little difference, apart from the faster growth of the neck and thoracic muscles in boars; this, of course, is analogous to that found in bulls and, to a lesser extent, in steers. Barrows were found to be very little different to gilts in this species.

The interesting difference is that the pig has a much greater proportion of the total muscle weight surrounding the spinal column, where there are some of the most desired and hence expensive cuts. It seems less likely that this occurs because of selection for what is obviously a valuable trait, than because it is a common character of smaller species, since it occurs in the sheep and deer. The muscles of the lower parts of the limbs in the pig are comparatively undeveloped, presumably a reflection of the low level of agility of the species.

In sheep

Recent studies have shown that breed has no effect on the relationship of the weight of the individual muscles to total muscle weight. There is the similarity with cattle breeds that muscle groups of an early maturing breed, like the Southdown, will undergo relative change in development at a younger age than a later maturing breed, such as the Romney Marsh. Again, there is the more marked development of loin muscles in sheep compared to cattle, as is found in pigs.

Carcass Assessment

It is the aim of all breeding schemes, feeding regimes and management systems concerned with meat animals to satisfy a consumer demand that has always varied in minor ways. This is most obvious in the way that carcasses are appraised, cut up and sold. In the past, this has not only varied between countries, but even between different parts of the same state, some communities liking more fat on their meat than others.

Today there is a marked increase in internal and international trade in meat, now becoming less with whole carcasses more in pre-packed form. This is allied to a requirement for less fat with all cuts. There is clearly a need to be able accurately and speedily to measure the amount of lean meat in a carcass.

For the farmer, as for the research worker, it would be useful to know the muscle weight of the live animal at any particular stage of its growth. This can at present be undertaken by the use of several methods. These include the whole-body counting of an injected dose of radioactive potassium, by the use of tritium to discover the amount of body water, and the use of harmless but biochemically detectable dyes for the same purpose. While acceptably accurate, these and other methods so far developed are too slow and expensive for routine use.

Important for the farmer, but particularly for the butcher or processor about to purchase a carcass, is the need to be able to obtain a rapid and accurate assessment of the total quantity of fat and lean meat available for sale.

Many methods are at present being used, some visual, others measure the backfat thickness or the depth of the 'eye' muscle because of its weak correlation to total muscle weight. Improvements to the accuracy of indirect measurements continue to be made and, in the meantime, the information coming from the grader at the point of slaughter remains the practical point of reference for the individual farmer.

6

Reproduction

Much of commercial animal production depends upon reproduction processes, for example, egg-laying in poultry, milk production in cows, sheep and goats. Selection for the development of these economically important traits has been undertaken over a comparatively long time. The speed with which improvement can be made or selection pressure applied has been improved by modern animal breeding techniques, amongst which is artificial insemination, record keeping on the farm, the use of computers and ease of communication.

Like any other character on which improvement is concentrated over many generations, there is the likelihood of a plateau being attained. So far, increases in two important characters associated with reproduction in the dairy cow and hen are still occurring.

It has been established that characters associated with reproduction have a low heritability. Most estimates place the heritabilities of milk production in cattle at 0.3 and of egg production in fowls at between 0.3 and 0.4. Too much must not be made of heritability estimates because their accuracy is directly related to the number of records examined and has only complete validity for those animals at that time.

If reproductive processes cannot be easily improved by the technique of selection, it emphasizes the importance of understanding the physiological processes involved. This permits us to devise routines of careful husbandry and, with the employment of the full resources of preventive medicine, to make sure that the genetic potential is fully realized.

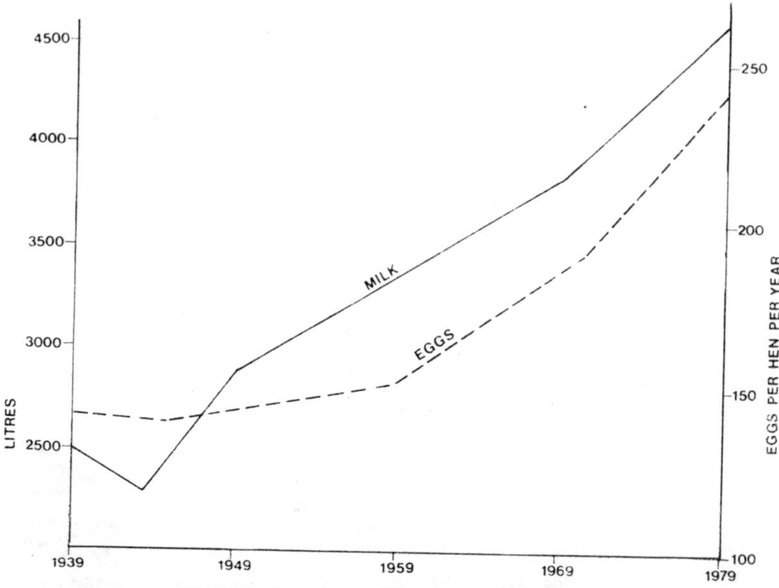

Fig. 6.1. Changes in milk and egg production.

Comparative Physiology of Reproductive Processes

The domestic and companion animals embrace two of the main divisions into which higher forms of life are divided, mammals and birds, each having different reproductive systems. In mammals, from the smallest per mouse to the elephant, the developing embryo remains within the uterus of the female until it has reached a relatively advanced stage of development, when it is able to walk or crawl round to the udder of its dam and suckle. The hen, after a successful mating, lays a fertilized egg that has to be incubated, that is, kept at a suitable temperature and humidity, and regularly turned. When the developing embryo, the chick, has reached a sufficiently advanced stage of development, it pecks its way out of the egg-shell to begin its independent existence.

Male

Testes

The genetic material of the male of every species, that is to say the genes being contributed to the next generation, are carried in the spermatozoa or sperms, produced in large numbers in the two testes which in all mammals are of equal size. In birds, only the right testis is functional. In order that there may be the same number of

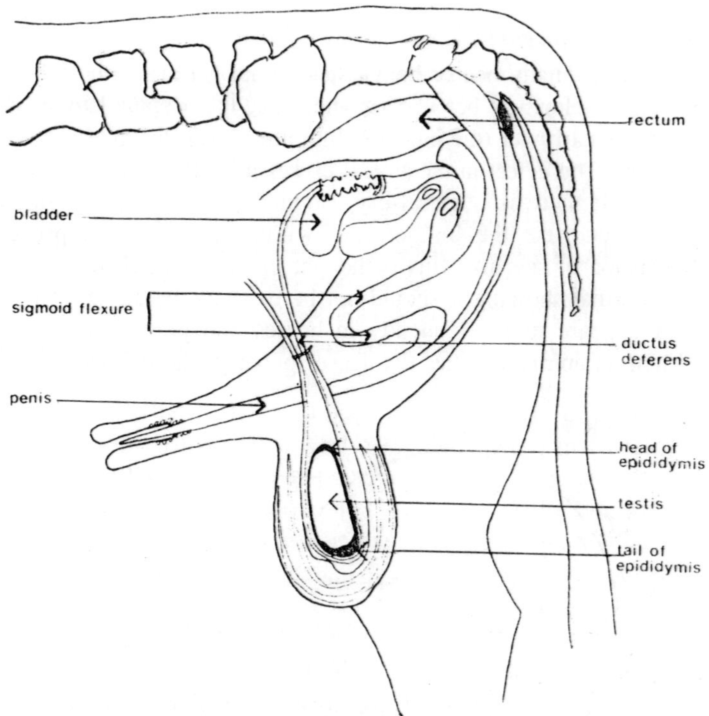

Fig. 6.2. Anatomy of reproductive organs of the bull.

chromosomes in the new individual that begins its life after the sperm has penetrated the ovum, there is a process of halving of the number of chromosomes of the developing sex cells of both male and female.

There are minor variations in the anatomy of the male reproductive organs of the various species.

Position of the testes

The two testes are held in the scrotum, the small skin-covered sac between the thighs of most animals. The boar and the tom-cat have the scrotum placed posterior below the anus. Some have the testes lying horizontally in the scrotum (stallion), most lie vertically; but all are outside the body. This is because the production of spermatozoa and their storage can only proceed normally at a temperature of several degrees below that of the body. One can notice that during hot weather, the muscles holding the scrotum relax so that the testes are held at a lower level. During very cold weather, it has been found that the internal scrotal temperature is between 3 and 5°C less than the body temperature.

Rig or cryptorchid

Occasionally one or both testes do not get drawn into the scrotum during development before birth and, since the opening into the abdomen becomes narrower (the inguinal ring) after birth, the animal has reduced fertility. The defect in the animal causes it to be known as a cryptorchid or rig. If there is bilateral cryptorchidism, the animal is likely to be sterile because the temperature within the abdomen prevents the formation of sperms. This is thought to be an inherited defect, so that any animal showing it should not be used for breeding. It is a defect that appears more frequently in horses and pigs than in ruminants. When an animal reaches the age of puberty, the process of producing

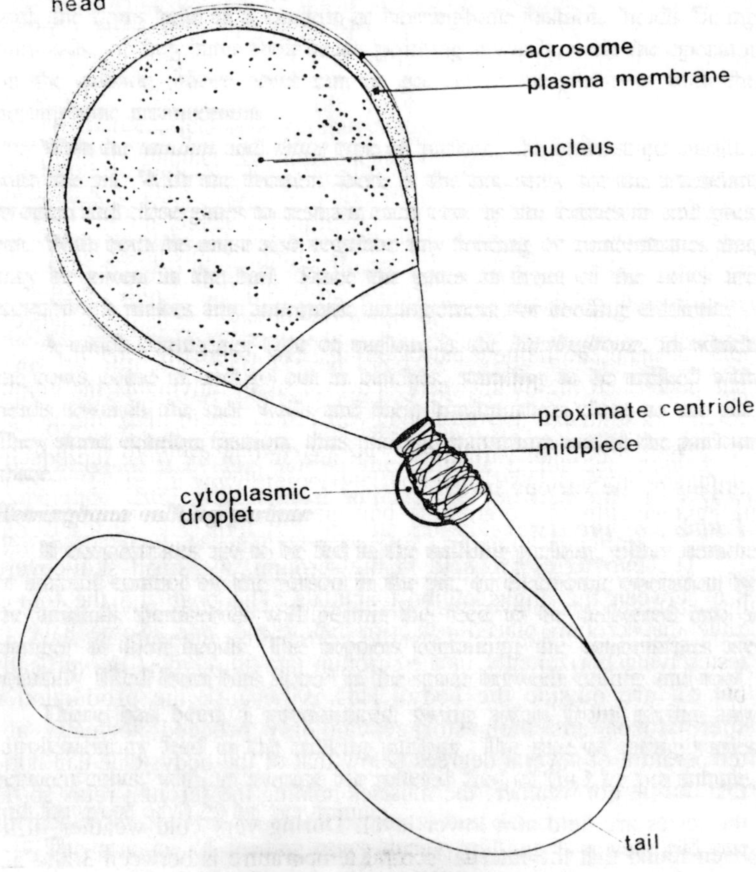

Fig. 6.3. Mammalian sperm.

the sperms takes place in the substance of the testis from special cells, the spermatagonia, lining the closely packed seminiferous tubules. From here they pass into the long coiled tube of the epididymus (in the bull, this can be up to about 150 m in length). Entering at the top, or head of the epididymus, the sperms steadily get moved along by the contractions of its thin muscular walls and the action of its own movement, thus gradually progressing through the central section, closely attached to the testis, and along the protruding tail at the other end. From here they enter the larger tube, the vas deferens, and ascend the spermatic cord into the abdomen; finally they enter the urethra, from which they may be ejaculated from the penis.

The sperm reach maturity as they travel along the epididymus. Within the abdomen are situated the accessory sex organs, the seminal vesicles, prostate and Cowper or bulbo-urethral glands. They discharge their contents into the urethra, the tube that carries the sperms to the exterior. These glands vary in size and activity with different species, e.g. in the boar the bulbo-urethral glands are large, but are absent in the dog. In the fluid of their secretions are salts, buffer and sugar to activate and nourish the rapidly moving sperms, as well as providing additional fluidity.

Each sperm consists of a head, in which is contained the chromosomes within a relatively large nucleus, covered by the acrosome, or cap. The small-mid-piece ends in the long tail, whose lashings help to propel it forwards.

The vas deferens enters the urethra at the spot where this tube comes from the neck of the urinary bladder. It is here that the secretions of the accessory sex organs mix with the sperm to produce the semen.

Penis

This is the copulatory organ and has important modifications in shape. In all animals it lies within the protection of the sheath, attached to the underside of the abdomen. This is lined with a membrane in which there are secretory glands, and which is continuous over the penis itself. That of the stallion has an enlarged end, with the whole organ provided with a plentiful blood supply. When sexually aroused, the blood supply is greatly increased, causing enlargement and protrusion from the sheath. In ruminants, there is an 'S' bend or sigmoid flexure which straightens during mating to permit the penis to emerge from the sheath. The ram and the goat have a thin urethral process at the tip of the penis. At joining with the ewe, this rotates within her vagina, but is coiled in the billy-goat. The boar's penis has a twisted

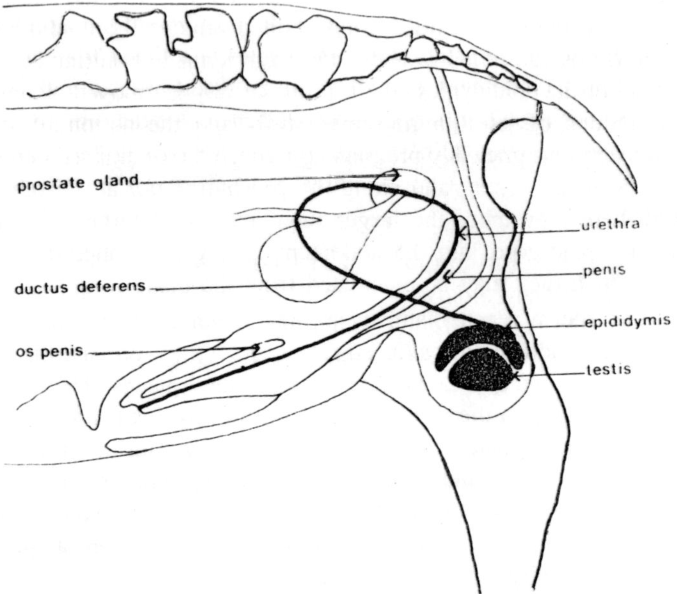

Fig. 6.4. Anatomy of the reproductive organs of the dog.

or corkscrew glans at its end; while in the dog there is a thin bone towards the tip, grooved underneath to take the urethra. It also has an enlargement behind it that swells during copulation, tightly gripping the interior surface of the bitch's vagina. The tom-cat has small projections pointing backwards.

Male mating behaviour

All the male farm and companion animals show characteristic behaviour at mating, that is to say, when they are brought into contact with a receptive female. To be willing to receive the male, she must be at a particular stage of her oestrous cycle during which she will show various signs of 'heat' or oestrus.

With the horse, a species in which hand or controlled mating is the common system, the mare thought to be in oestrus is brought into the presence of the stallion. If this procedure is attempted when the mare is not on heat, the mare may violently repulse him. Thus, when the mare is brought to the stallion, spoken of as a 'teasing', it is important to have solid fences or a gate between them. Should the mare be receptive, she will stand quietly and mating can safely follow.

The stallion and the bull show characteristic upward turning of the head, curving of the upper lip and flaring of the nostrils, spoken of as the 'flehmen'. Particular odours are associated with the different sexes, known as pheromones. Those produced by the female in oestrus help in her identification by the male; those produced by the male help to provoke receptivity in the female. With sheep and with pigs and cattle under some systems of management, where mating is uncontrolled, the identifying pheromones play an important part in the mating process.

Disorders of Reproduction in the Male

There can be a number of reasons for a male animal being incapable of producing normal semen, or there may be disease or injury involving the external genitalia that may interfere with his capacity or willingness to mate.

Testes

Injury may cause inflammation and swelling of one or both testes, a condition known as orchitis. An infection of the skin of the scrotum can also affect fertility. Any general infection that causes a raised temperature will inevitably cause a temporary sterility. There may be infection of either or both testes from blood-borne organisms, and routine examination by palpation should be made before purchasing any male animal or before the beginning of a mating season.

Epididymitis

Infection and enlargement of the epididymus, particularly in the ram, is by no means uncommon. In Australia, New Zealand and parts of Latin America, infection with *Brucella ovis* has been a problem, although several other species of bacteria can also cause inflammation in the epididymus. There is initial swelling, most marked in the protruding 'tail', easily felt at the lower end of the testis.

It is a wise precaution to make a manual examination of a bull's testes before using him. Carefully and gently done, it can detect changes in the organ itself and in the associated epididymus. The normal gland feels smooth and uniformly tense but not hard.

Method of Examination

If a ram is being examined, he should be up-ended and held, sitting on his rump, by an assistant. The testes should be gently taken to lie in the palms of the hands, and the substance of each is felt to locate any area of inflammation, causing increased tenseness. A hardness will be felt if there has been fibrous tissue laid down after infection. The thumb and forefinger can then proceed to feel the epididymus, starting at the head and going down to the tail, before going upwards to the vas deferens part of the spermatic cord. There should be no thickening here or break as it mounts towards the inguinal ring, through which it passes into the abdomen.

The sheath, lined with the prepuce, can become infected, which may follow partial blockage of the urethra from calculi or small stones formed in the urine. The inflammation caused by this state of affairs has been called a balanoposthitis. It is a condition more commonly seen in wethers than in rams, goats or bulls. It demands veterinary treatment without delay. Bulls may be affected by a somewhat similar condition in which the prepuce prolapses, that is, it becomes slack, not being held in place by the muscles that normally do so. The tip of the penis is frequently seen protruding from the everted prepuce. Amongst bulls, it is often those of the beef breed that show this disorder when out at pasture, being especially common in zebu breeds, whose sheaths are always more pendulous. The treatment consists of clipping away the long hairs around the sheath and then carefully washing the part in warm, soapy water. Using more warm water, gently massage the tissues to increase the circulation. After drying, place some mild disinfectant ointment on the part. After repeating this procedure several times, it should become possible entirely to replace the everted tissue and return the penis to its normal position in the sheath.

More rarely, there may be injury to the penis, or inability to withdraw the penis from the sheath because of adhesions. Rest alone may bring about recovery, otherwise investigation under anaesthesia becomes necessary.

Occasionally, a male animal of reproductive age will show an unwillingness to mate with receptive females. This failure of libido can have several causes, apart from the effects of systemic or localized infection. Young rams need to be mixed with older fertile rams to encourage sexual activity. Beef bulls are often slow to begin service after reaching puberty at about 10 months and lose sex drive before dairy bulls, at about 6 years.

Bulls that have been successfully used in artificial insemination centres may suddenly refuse to serve or there may be failure of erection. This may be due to some accidental painful experience during service, but often the cause is difficult to ascertain. The use of a new 'teaser' cow of different appearance, or by simply resting the bull for a month or two may restore sexual vigour.

It is perhaps not surprising that the sperm cell concentration falls and the proportion of sperm showing defects rises during the hottest months when bulls of the European breeds are used in tropical countries. One recent study in West Africa showed that with lower temperatures during the wet season, there was an improvement in sperm density and sperm morphology.

ARTIFICIAL INSEMINATION

The technique of collecting semen from particularly desirable male animals of the various breeds of farm and companion animals and transferring this to a receptive female of the same species, that is, inseminating her, is one that is widely used, particularly in the dairy industry. Improved methods of diluting and storing semen, especially of the bull, has made it possible to inseminate from 40,000 to 60,000 cows per year from a single bull. It has been pointed out that it would be possible for 100 superior bulls to inseminate all the cattle in the U.K.

Collection of Semen

There are three methods for collecting semen. The commonest is to use an artificial vagina, consisting of a double-walled cylinder into the interior of which the male thrusts his penis and the outer chamber, filled with water at blood heat simulates the female vagina. The ejaculated semen is collected in a glass tube at the opposite end. The artificial vagina is made of a hardened rubber outer cylinder and the interior consists of a thin rubber tube inverted over the outer case which has an opening into which warm water is poured to give an interior temperature of 45-55°C. The male is lead to a female in oestrus ('on heat') and as he mounts her, the penis is directed into the artificial vagina. At Artificial Insemination Centres for dairy and beef bulls, it is often not possible to have a cow in season (in oestrus) available. Well-trained bulls can be satisfactorily stimulated by the presence of a quite cow not on heat, and at many centres the use of a 'dummy' consisting of a stretched hide over a frame is used. This can permit collection without diversion of the bull's penis by the operator crouching inside the frame. Stallions normally need the stimulus of a mare in season.

The second method, used more often with the ram and goat but sometimes with old lame bulls, is to use an *electro-ejaculator*. Placed in the rectum, this passes an electric current through the accessory sex glands and, as the stimulation proceeds, semen is produced and collected.

The third method is not used as a routine measure but can be used if a sample of semen is required and a bull has not been trained to serve into an artificial vagina or no cow in oestrus is available. A gloved and lubricated hand is put into the rectum and the seminal vesicles are gently stroked against the pelvis. The ampullae are then massaged and the semen that begins to drip from the sheath is caught in a glass funnel fitted into a glass tube held under the sheath opening. If semen is to be collected from a stallion, a condom may be used.

The glass tube into which the semen is collected must be warm, close to body heat, to prevent cold shock. It must at once be placed in warm water to maintain this temperature unless the air temperature is high.

Evaluation of Semen

Before semen is used it is important to have an assessment of its quality. An approximate estimation can be made by rapid inspection of the whole ejaculate. It should appear uniformly milky with no extraneous material such as blood or pus. Routinely, at Artificial Insemination Centres, a sample is immediately taken and a sperm count made. The quantity of semen in an ejaculate and the density of sperms in relation to the accessory glandular secretions vary considerably from species to species and even within a species, individual animals may vary widely in their characteristic sperm density. The extent of this variation is seen from the generally accepted values.

Table 6.1. Characteristic sperm densities

Animal	*Volume ml/ejaculate*	*Sperm concentration million/ml*
Horse	60-100	100-150
Bull	0.5-12	1000-1800
Boar	200-300	200-500
Ram	0.8-1.2	2000-3000
Dog	3-25	4-540
Cock	0.5-2.0	800-6800

In normal semen, not only will the sperm count be within a stated range, but the sperm will show great activity and there will be no more than 10 per cent with any abnormality, e.g. lacking the acrosome or with broken, bent or incomplete tails.

Dilution of Semen

Normally, in all species, following checking on density and viability, semen is diluted by mixing it with diluting fluid the composition of which affords nourishment to the sperms, buffers the mixture to prevent accumulation of acid and contains antibiotics to counter bacterial action.

The amount of dilution depends upon the species, but, as an illustration, bull semen that is to be kept in the deep frozen state is diluted 10 times with an egg-yolk citrate or skim-milk extender. This commonly takes place at 25°C, before cooling to 5°C, when an equal volume of diluent containing glycerol is slowly added. Fresh bull semen has an approximate concentration of 1000 sperm per ml and it is accepted that an effective fertilizing dose is between 5 and 12.8 million. It means that one ejaculation from a bull is sufficient theoretically to inseminate 1000 cows.

Deep-frozen Bull Semen

For many years following the discovery of the protective effect of glycerol, semen from bulls used to be frozen at –79°C and, so long as there was no rise to –69°C, there was only slow deterioration in fertilizing ability. Now the use of liquid nitrogen allows semen to be kept at –196°C. The form in which the semen is stored and the manner of its use have also changed. From storing in small glass ampoules, the common methods at present are either to put the individual dose into a plastic 'straw', a thin tube of polyvinyl with 0.25-1 ml capacity, or by pellets, the method favoured in the U.S.A. These are formed by placing a drop containing 0.075-0.1 ml of diluted semen into holes in a block of solid CO_2, the pellet so formed representing one dose containing 12-30 million sperms. This is stored in liquid nitrogen wrapped in foil or in storage tubes.

Insemination

A successful insemination depends upon three factors, the recipient female, the skill of the inseminator and the quality of the semen.

Whatever the species of female animal that is to be inseminated, all tend to release a ripened egg from the follicle in the ovary (ovulation) towards the end of the period of oestrus or soon after its ending. This indicates the best time to inseminate.

Mares

Although not widely used, except within large horse studs, the steady increase in the number of light horses being bred all over the world for recreational purposes has stimulated research into making deep-frozen semen available. The difficulty has been in finding a satisfactory protective dilutent, since glycerol tends to depress fertilizing ability. Larger numbers of spermatozoa per insemination is used to compensate. It has also been found by workers in North America that there is a substantial improvement in the volume of ejaculate according to the season, with an increase of about 38 per cent from winter to spring, with a higher density of sperms. Fertility with frozen semen is mainly within the range 50-65 per cent.

With fresh semen, without dilution, and using 10-50 ml for each insemination (containing 1-2 billion sperms), four mares could be inseminated from one ejaculation. The mare is held in a crush and with the tail bandaged and the vulva washed. The sterile insemination catheter is guided into the dilated cervical canal by a gloved and lubricated hand inserted into the vagina. The fertility from using fresh semen is comparable to that from natural service.

Cows

The oestrous cycle of European breeds of cattle has an average length of 21 days, with greater variation being shown by zebu breeds as well as a shorter time during which signs of heat are shown. Ovulation takes place at about 12 hours after signs of oestrus have ceased, but maximum fertility results from insemination 6-24 hours before this.

Frozen semen is thawed just before being used and, by means of the appropriate instrument, the semen is inserted into the front part of the cervix (the short connection between the uterus and the vagina) or it is put just into the uterus. To do this, the inseminator places one hand into the rectum to hold the cervix and bring it slightly forward, thus to guide the sterile plastic pipette or the plastic 'straw' to the internal opening of the cervix. The dose of semen is injected by means of a syringe if the pellet form of semen is used, or by the plunger of the special applicator if a 'straw' is used, following the removal of the plugged end.

Assessment of fertility is by calculating the percentage of cows holding to first service (i.e. first insemination). If a cow does not return to oestrus during the following 60 days, it is likely that she has conceived. It was wise practice to have confirmation of the cow's pregnancy, either by means of a rectal examination by a veterinarian, or by ascertaining the level of progesterone in the milk. The Milk Marketing Board in the United Kingdom began to offer the latter service to dairy farmers when the technique of detecting the raised level following conception became reliable and could be automated.

Continuing shortage of dairy products in tropical countries has encouraged the use of frozen semen from high-producing progeny-tested bulls of European dairy cattle for improving yields by cross-breeding with zebu breeds where the standard of nutrition and general management of herds of tropical dairy cattle has been sufficiently improved to allow practical utilization of the higher genetic potential.

The use of performance tested beef bulls for artificial insemination has not been so popular, although such services are widely available. The main reason is that beef cattle are most commonly kept under extensive systems of husbandry, allowing calves to run with their dams over relatively large areas. Thus there are difficulties in detecting cows in oestrus and arranging to have semen brought at the right time. An effective way of overcoming this difficulty is to use 'teaser' or vasectomized bulls which can have a 'chin-ball' attached to a head collar which will mark cows found by the teaser to be on heat. Another device is to use adhesive Kamar pads on the rumps of cows so that, when they come into oestrus and stand to be mounted by another cow, a clearly detected dye stains the hindquarters.

Sheep

Except in the Soviet Union and some eastern European countries, where problems of cost-effectiveness are not so closely considered and where there has been the need to rapidly improve production characters, the use of artificial insemination in sheep has not become widely used. For many years, freshly collected and diluted semen was used. The technique of deep freezing ram semen in pellet form is now available, but it is widely held that the intense selection pressure long exercised in Australia for superior wool quality, as well as the availability of rams of the meat-producing breeds, considered in relation to the comparative low value of individual ewes, makes an insemination service uneconomic.

The procedure of inseminating a ewe, found to be on heat by the use of a vasectomized ram wearing a marking harness, is to have her held in a crate or cradle that presents the hindquarters to the inseminator. The vagina is opened by a perspex cylinder, the interior illuminated and the pipette guided into the cervix.

Sows

Many European countries and part of the United States, where pig production is an important livestock industry, have identified superior boars through progeny testing and performance testing schemes. When an artificial insemination service is available, many small producers find it uneconomic to keep a boar, and large enterprises that are closed because of the imposition of strict disease control, use it to introduce new genetic material. The detection of oestrus in sows when no boar is present is by no means easy. There are certain indications, such as a degree of restlessness in the sow, some swelling and reddening of the vulva; but the best indication of true heat is the *standing test* when the sow stands still and permits an attendant to put his weight on her hindquarters. With gilts, however, up to one-third may not react to this test. It is more effective to allow such young females to be brought within the proximity of a boar when the smell of the pheromones from his salivary glands and the sounds he makes will provoke stronger signs of oestrus.

Because of the circular folding of the cervix in the sow, the rubber catheter used is also spiralled at its tip, thus locking into place. It has been the practice to make up an inseminating dose to a volume of 50 ml containing 2 billion sperms. This is slowly transferred over 8-10 minutes under gravity into the uterus.

At most centres where boar semen is collected, the use of an artificial vagina has been discarded in favour of the gloved hand method. Most boars, after training, will serve on a dummy sow. This consists of a padded metal frame with a canvas cover. The gloved hand of the collector firmly grasps the penis, with its ridges held between the fingers. The first part of the ejaculate, the accessary fluid, is not collected. In contrast to the swift ejaculation of the bull and stallion, it takes about 10 minutes to collect the sperm-rich and subsequent parts of the boar's ejaculate (100-350 ml).

Female

The contribution made by the female to her offspring is two-fold. Firstly, she makes an equal genetic contribution as the male, passing half her genes in the ovum (egg), released at the end of the oestrous cycle from the ripened follicle on the surface of the ovary. Secondly, as a result of the environment that she provides within the uterus during the time of gestation of the developing offspring, many opportunities occur for things to go wrong; for example, nutritional deficiencies in the diet or infection by pathogens.

Ovary

All farm and companion animals, except the fowl, have two ovaries located below the kidneys on both sides of the abdomen.

In the fowl, only the right ovary becomes functional. It is during the early cell divisions of the developing embryo within the uterus, that the first stage of meiosis occurs, but the second stage, which includes the actual formation of the ovum, only occurs as it reaches maturity during each oestral cycle. In the multi-ovular species, such as the pig, dog and cat, a number of eggs reach maturity and are released into the oviduct at the same time.

Oestrous Cycle

This is the term applied to the sequence of events that occur between the release of ova. Some species cycle throughout the year, others are seasonally polyoestrous.

The normal cycle can be divided into five parts. Just before oestrus there is a short pro-oestrous period when the ripening follicle in the ovary rapidly increases in size. The uterus enlarges and its serous lining thickens with its many glands becoming active. The lips of the vulva enlarge and, in ruminants, there is sometimes a slight discharge of clear mucus. In the bitch this discharge is blood stained.

Table 6.2. The oestrous cycle

Animal	*Length days*	*Time of year*
Mare	21 (19-26)	Spring and summer
Cow	21 (18-24)	Throughout year
Ewe	17 (14-20)	Autumn; Dorset horn and Merino and tropical breeds cycle throughout year
Goat	18 (18-21)	Autumn
Sow	21 (16-24)	Throughout year
Bitch	21 twice yearly	January-March; August-September
Cat	3-4 weeks	Spring; late summer; autumn

Table 6.3. Heat period and ovulation

Animal	*Duration of heat*	*Ovulation*
Mare	6 days	24 hours before the end of oestrus
Cow		
European	ca. 24 hours	14 hours after the end of oestrus
Tropical	3-7 hours	
Ewe		
British breeds	36 hours	Near the end of oestrus
Merino	48 hours	
Goat	24-48 hours	Near the end of oestrus
Saw	40-60 hours	30-40 hours after beginning of oestrus
Bitch	9 days	5-12 hours after the end of oestrus
Cat	3-6 days	26 hours after mating
	5-12 days if not mated	

During oestrus itself the vagina enlarges and the cervix relaxes. This is the time that the females of all species seek to mate with the male and will stand to do so. The time that this lasts varies, not only between species, but between breeds and individuals.

When the period of heat ends it is succeeded by metoestrus, during which the ruptured follicle in the ovary fills with luteal tissue to form the corpus luteum, or yellow body. Dioestrus is the time that elapses before the beginning of the next oestrus. When an animal has a period of sexual rest she is said to be 'in anoestrus'. Now the ovaries are inactive and the uterus contracted. Sheep, goats, dogs and cats, especially in temperate regions, tend to have anoestrous periods lasting several months.

Pregnancy

The egg, after its release from the ovary, enters the fallopian tube through the distal expanded end and begins its descent towards one of the two horns of the uterus with which it connects. If there has been a normal mating, or the introduction of spermatozoa by artificial insemination, the upward swimming sperms very quickly reach the upper part of the fallopian tube. One sperm gains entry, although it is accepted that up to a million should be introduced when using artificial insemination. The first mitotic division immediately takes place. Further division occur in what is now known as the blastocyst as it enters the uterine horn and becomes attached to the lining membrane after several more days, the time taken varying from 12 days in the cow to 30 days in the mare.

As the embryo develops and a normal pregnancy proceeds, the developing embryo is enclosed within a thin-walled sac, the amnion, which itself is within and connected by the umbilical cord to another sac, the allantochorion. The latter becomes attached to the lining of the uterus. The form of this attachment varies with the species. In ruminants, there are a series of protuberances, the *cotyledons* or caruncles, about 120 in the cow and 80 in the ewe, arranged in four rows along each of the uterine horns. These provide the means for the dam's circulation to nourish the new offspring, now known as the foetus.

There is a different arrangement in the pregnant mare, sow, bitch and cat. These have a large number of small finger-like processes called *villi* making the connection between the allantochorion and the endometrium or lining of the wall of the uterus. In the mare and sow these are found all over the outer surfaces of the foetal sac, while in the bitch and cat it is only the central part that becomes vascularized (having a supply of blood from the mother). It is this foetal sac that is known in all mammals as the *placenta*.

Throughout its intra-uterine life, the foetus floats in the amniotic fluid which, in its turn, is surrounded by increasing amounts of allantoic fluid, together making up the foetal fluid. This may be as much as 20 litres at the end of the pregnancy in the cow.

Factors Controlling Fertility

It is a prime aim with farm livestock that young females should be fit to mate at as early an age as possible to permit the production of vigorous offspring of optimum size. A long life of regular breeding should follow such a rearing routine, assuming freedom from disease and a continuation of good husbandry. It requires careful attention to the feeding and management of the prospective dam from its birth.

Consider the growing female calf, the heifer, as an illustration of what should be done. There is a direct relationship between the level of nutrition given to a heifer calf and the age of puberty. Thus it is possible, by persistent high feeding from an early age, to obtain a successful mating when a calf is less than 15 months old. The first lactation yield of milk from such 'forced' females is almost always disappointingly low. Maximum lifetime production from cows is more likely to follow a rearing period, during which the female calf is fed a carefully balanced diet at a moderate level from weaning onwards and is mated for the first time at about 18 months of age.

With all the common domestic animals, the arrival of puberty comes before growth is complete. Fertility is likely to be low and uncertain, so the time of first mating is delayed to an extent that varies with the species.

Table 6.4. Age of puberty and mating

Animal	*Puberty*		*Mating*
	Mean, months	*Range, months*	
Mare	18	12-24	3 years
Cow	13	6-12	18 months
Ewe	10	6-12	18 months
Goat	7	5-9	18 months
Sow	7	6-8	8 months
Bitch	9	6-12	12 months
Cat	6	5-8	10 months
Rabbit	6	18	
Fowl	22 weeks	18-24 weeks	over 1 year

The filly foal will reach puberty at an age that is even further from the time that growth ceases, i.e. the age of maturity. It is therefore desirable to delay first mating until nearer that stage, if growth is to be uninterrupted and difficulties at parturition are to be avoided.

With modern strains of crossbred ewe lambs, it has become increasingly common, on properties able to provide optimum nutrition during their first winter and good care at lambing, to undertake their mating at about 8 months. Fertility is below that obtained with mature ewes, but subsequent reproductive capacity does not appear to be adversely affected.

The endeavour with modern strains of gilts is similar. Many breeders may still be content with mating at slightly over 9 months at a weight of 130 kg, but it is the aim of progressive farmers to mate gilts two months earlier when their weight is likely to be close to 105 kg. A satisfactory breeding life and optimum litter size depends upon adequate feeding of a balanced diet, not only through each pregnancy, but during lactation and the dry period that follows.

Control of Oestrus by Hormones

A series of chemically identified and naturally produced hormones regulates the succession of events that have already been described as making up the oestrous cycle. They have three sources: from that part of the central nervous system called the hypothalamus, from the nearby anterior pituitary gland (both situated at the base of the brain), and from the ovary and uterus. It appears that with most species, there is an external stimulus necessary to initiate regular heat periods.

With small ruminants and the mare, it is the seasonal change in the day length that initiates the oestrous cycles after puberty, assuming normal temperature and optimum nutrition. With the mare, it is the increasing length of the daylight in the spring and, with the ewe, the decreasing daylight of autumn that has this effect.

The hypothalamus produces certain 'releasing factors' which act on cells in the nearby anterior pituitary gland to produce *follicle-stimulating hormone* and *luteinizing-hormone*. The action of the former is to stimulate the cells within the ovary to proceed to the ripening of the follicle, simultaneously bringing about a release of *oestrogen* into the circulation. Its main effect is to cause the various observable signs of heat.

It is immediately followed by production of luteinizing hormone, causing the cells in the ovary to secrete increasing amounts of *progesterone*. This is accompanied by the enlargement of the corpus luteum in place of the ruptured follicle. The effect of progesterone is to abolish all signs of oestrus, bringing about the state of anoestrus. If pregnancy follows, the level of progesterone in the circulation (and in the milk of a lactating cow) continues to rise until about the 20th day, but stays at a fairly high level until after the 40th day in this species. The placenta also produces progesterone as a function of maintaining pregnancy.

Apart from the central nervous control by the hypothalamus in producing successive releasing factors to the anterior pituitary, there is a powerful group of hormones, the prostaglandins, produced in various tissues; but one, prostaglandin $F_2\alpha$, from the uterus has the effect of causing the destruction of the corpus luteum, and thus the rapid diminution of circulating progesterone. In the non-pregnant animal this signifies the end of the dioestrous period of the normal oestrous cycle and the beginning of pro-oestrus. In the pregnant female it occurs when parturition or the birth process is coming close.

Detection of Pregnancy

It is desirable to be able to be sure that an animal is indeed pregnant at as early a time as possible. In some species, such as the mare, bitch and cat, where each individual is often under close observation, the missing of the succeeding heat period after mating is a hopeful indication of pregnancy. With herd animals, cows, sheep, goats and sows, such close observation may be difficult, although the keeping of records specifying the date on which mating began and the employment of marking techniques can provide information about the commencement of parturition.

Mare

Two methods of detecting pregnancy are commonly used. The most reliable is for a veterinarian to make a clinical examination. This may be done in one or both of the following two ways: a rectal examination is undertaken, during which the uterus is felt with the hand and arm inserted into the rectum. The best time for this to be undertaken is 35-40 days after mating. Sometimes an examination of the vagina is made using a speculum, for there are a series of changes in the lining cells, the amount of mucus present and the appearance of the external opening of the cervix.

The second method is to take a blood sample to see if there is present the special hormone produced in large quantities by the pregnant uterus. It is called 'pregnant mare serum gonadotrophin' and has both oestrogenic or follicle-stimulating and luteinizing properties. It is first demonstrable in the blood serum of the mare between days 38 and 42, reaches a maximum at about 60-65 days and disappears by the 150th day. A blood sample should therefore be taken between days 60 and 100. In the *mouse test* a small amount of the serum is injected twice daily for two days into each of four female mice 3-4 weeks old. When the mice are autopsied some 24 hours later, the test is positive if the uteri are some five times normal size.

In the *haemagglutinin inhibition test*, a sample of the mare's blood serum is diluted with acetone, and the resulting precipitate put into buffered saline to be incubated for one hour. Specially prepared sensitized sheep erythrocytes and anti-pregnant mare serum are mixed with increasing dilutions of the mare's serum. If the mare is pregnant, there will be no clumping together of the red blood cells.

This test has about a 90 per cent level of accuracy and is believed by some experts to be preferable to making repeated clinical examinations. These have become routine in many horse studs, but there is the possibility that the low levels of fertility commonly experienced amongst thoroughbreds in many countries may be in part due to this manual interference.

Cow

Until the mid-1970s, the most reliable method of ascertaining whether a cow previously mated, and not since seen to show oestrus, was already pregnant or barren was to have a clinical examination made per rectum. The various changes that take place in the size and position of the uterus of the pregnant cow can be most certainly detected when an examination is made after 90 days, but experienced veterinary practitioners can make accurate assessment a month or six weeks before this.

With the greatly increased accuracy that the use of a radio-immune assay for progesterone in the milk gives, especially when it was found possible to automate the procedure, it has become possible for organizations, such as the Milk Marketing Board of England and Wales, to offer their members a reliable diagnostic service. A sample of milk is taken from the cow between the 18th and 24th day following mating or artificial insemination and sent to the Milk Marketing Board laboratory. The result is sent back within a few days. Accuracy is over 90 per cent with negative results and a little over 80 per cent for positives.

Progressive dairy farmers, anxious to maintain high levels of production by endeavouring to have their cows produce a calf each year, use both aids to discover if mating or artificial insemination has been successful. A more recent development has been the discovery that some cows always have higher than normal progesterone levels. These would account for the few false positive readings from the milk test, discovered later by the *per rectum* examination. A blood sample taken at the time of examination will detect such cows, for at this time progesterone levels are normally low.

Ewe and goat

A common and satisfactory way to monitor the occasion of the ewe being mated is to place a marking device on the ram. One widely used, first in Australia and New Zealand, but for many years by British sheep farmers, is the 'Sire-syn- harness. This has a socket in the harness that fits over the brisket, containing an easily changed block of raddle. If the colour is altered every 17 days, any ewe that does not conceive to a first service will have a second colour put upon her rump. For the valuable stud ewe, or nanny-goat, there are more sophisticated methods of pregnancy diagnosis available. These include using an ultrasonic foetal pulse detector. It has a probe that is moved over the bare or shaved surface of the abdomen immediately in front of the udder. Only when the pregnancy is over 60 days and after experience in its handling, is there likely to be an acceptable degree of accuracy. Beyond 70 days, radiography is very accurate but expensive.

Sow

Non-return to oestrus after 21 days and 42 days gives a clear indication that the sow is pregnant, but in large units where sows are often kept in stalls, signs of heat are often difficult to detect. There is available an ultrasonic foetal pulse detector in which a probe is placed on the abdomen or into the rectum. It works through the foetal pulse interfering with the transmitted signal. It is an expensive machine requiring a good deal of experience for effective use to be made of it.

There is also a method that requires the services of a laboratory. This involves the removal of a small piece of the lining of the vagina by a special instrument. A characteristic arrangement of the cells is shown by the pregnant sow, when the stained specimen is examined microscopically. The best time to test is between 18 and 25 days after mating.

Bitch and cat

Even if she is not mated, the bitch not infrequently has a false or pseudo-pregnancy. This state may persist for the whole of the 63 days that is the normal length of gestation, to end with the enlargement of the mammary glands and a behaviour pattern similar to the pregnant animal. A pregnant bitch tends to show behavioural changes, becoming quieter and more lethargic. There is a tendency to lay down subcutaneous fat and, if she is carrying many pups in utero, the abdomen will show distension by the 5th week. An infected uterus could be the cause of this, so it is better carefully to feel the uterus and the enlargements that are the growing foetuses. It is only easily done if she is not one of the large breeds and is of quiet disposition, so as not to resent this handling. It is most usefully done at about the fourth week of pregnancy. It should then be possible to feel the spherical swellings made by the foetuses along the horns of the uterus.

Because of the thinner adbominal wall in the cat, palpation is easier. If necessary, radiography can be used with a high degree of certainty after day 25 when ossification of bones has begun, and in skilled hands even earlier.

Parturition

The process of giving birth in mammals begins when the foetus has grown to a stage that permits it to fend for itself in the external environment. The procedure is controlled by a series of co-ordinated events, mediated through changing hormone levels in the circulation. These come from the central nervous system, the pituitary, the uterus, and from the corticosteroids produced in the adrenal glands of the dam and the foetus itself. It is important to be familiar with the normal length of the pregnancy, the gestation of each of the farm and domestic animals.

Table 6.5. Length of pregnancy

	Day	
Animal	*Mean*	*Average*
Mare	336	325-341
Cow: European	280	277-284
Zebu	286	284-288
Ewe	147	146-149
Goat	151	145-156
Sow	114	110-117
Camel	300	280-320
Buffalo	316	314-320
Bitch	63	60-66
Cat (queen)	63	60-66
Rabbit (doe)	31	30-33

A clear indication that birth is near is given by the enlargement of the mammary gland and teats of the dam. The lips of the vulva swell and may show some reddening, and there may be a slight haemorrhage. The ligaments attached to the sacrum relax in the cow, giving the hind quarters a sunken appearance.

The dam will almost always show behavioural changes as the period of gestation comes to an end. She will tend to go apart from the group and, if possible, seeks a quiet, secluded spot for giving birth.

When these premonitory signs occur, it is wise to bring the animals into special accommodation, a loose box for mares and cows which must be thoroughly clean and free from infection, well lighted and well ventilated, but without draughts. This accommodation should be in a quiet corner of the farm buildings, but of easy access to the animal attendant. It must have ample space without projections or encumbrances and thus should be larger than the usual stall or loose box. Plenty of clean bedding must be put down, a feeder at the right height from the ground, with drinking water always available. It is desirable for the parturient animal to have a day or two to become familiar with the place before the birth process starts. The nearness of parturition in the mare is shown by the enlargement of the mammary gland and a 'waxing' of the teats, sometimes with the escape of a few drops of colostrum. She may show patches of sweat by the elbows and flanks.

Normal Parturition

The effect of the changes in hormone secretion that occur when gestation comes to an end is to cause gradually increasing rhythmic contractions of the muscle fibres within the uterus. To begin with, in what is properly called the *first stage of labour*, these periods of contractions are not frequent and are interrupted by periods of rest. In the cow, these deep peristaltic waves that start at the distal end of the horn of the uterus and progress forward, occur every 15 minutes, become more frequent as the foetus is moved towards the cervix or vagina, until they occur every 2-3 minutes during the hour before the calf is born.

The cervix at the end of the vagina at this time becomes relaxed and enlarged to form a continuous canal, through which the foetus must pass, preceded by the allantochorionic sac.

It is at this stage that the foetus turns itself to be in the right position for being born. For the foal and puppy this means turning over with the head uppermost but, with the foal, calf and lamb, also an extension of the forelimbs. All foetuses at this stage extend their forelimbs forward, although many piglets are born with the hindlegs coming first. During this first stage, the dam is often restless, moving about or getting up and then lying down. It lasts from 3 to 6 hours, and by the time the second stage begins the dam lies to give birth.

The expulsion of the foetus or the second stage of labour is aided by contraction of the abdominal muscles in association with continuing uterine action. The allantochorionic sac breaks and much dark-coloured fluid comes from the vulva. As the head and shoulders of the foetus enters the cervical canal, a rise in oxytocin output from the posterior pituitary causes very strong muscular effort to force the foetus through to the outside, where it appears with the amnion or 'water-bag'. This is often broken by one of its feet or, with the mare, as it reaches the ground, or during the last part of its expulsion, and breathing starts. It is important to allow the umbilical cord connecting the newly born animal with the maternal membranes to break of its own accord following movement by the newly born animal or its dam as she gets up. This is important because blood-containing substances, that will give the young animal resistance to any pathogenic microorganisms in the surroundings, pass from the placenta at this moment.

This stage takes a varying time according to the species. It is shortest in the mare, where a mean time of 17 minutes is given and longest with the sow, who frequently takes over 4 hours before the last piglet is born.

The last or third stage of labour is simply the expulsion of the foetal membranes or afterbirth. It is swiftest in the mare, for it follows the birth of the foal within about 1 hour, but with the cow it is about 6 hours. All the farm and domestic species, except the mare, normally eat their afterbirth.

The dam, except the sow, will shortly begin licking her offspring. Partly through this stimulation and that of hunger, the young creature will find its way to its dam's mammary gland and begin suckling. The importance of receiving *colostrum*, the first secretion of the mammary glands, lies in its high nutritive value and in its concentration of protective antibodies, the immunoglobulins. The capacity of the intestine to absorb these proteins diminishes rapidly after 36 hours, which emphasizes the need for anyone attending animals at this time to see that the newly born gets its share of colostrum as quickly as possible.

Ewes

The lambing ewe flock is, in the U.K., increasingly being brought inside to lamb. The winter housing of ewes has been found to be economically justified under certain conditions of soil and climate, but in most sheep-farming areas a more limited type of under-cover accommodation, for different groups of ewes as they come to lamb, achieves the object of reducing perinatal losses, that is losses amongst both ewes and lambs that occur just before, at the time of parturition or during the first few days after birth.

It is an important reason for adopting the practice of using marked rams at mating. This makes it possible to draft ewes into groups according to the time of lambing, so enabling the vitally important extra feed to be given during the last 6 weeks of pregnancy and then to bring them into sheltered surroundings for the actual lambing. Whether this is entirely indoors or in an area protected from the worst of the weather outside, it must be clean and quiet. If outside; adequate protection can be provided by straw bales formed into small lambing pens. So long as the nutrition of the ewe has been correct, and measures taken to prevent disease, there will be little difficulty and few losses.

Sow

It is the commonest practice to-day to put sows about to farrow in one of the many varieties of farrowing crate. In this multiparous species, with litters of 8-12 usually being born, loss of piglets from the sow lying on them was a problem that the farrowing crate has largely solved.

It is placed in a pen that should be 1.5-2 m wide. Typical measurements are 2.11 m (7 ft) in length by 0.76 m (2 ft 5 in) wide and 0.88 m (2 ft 10 in) high.

The sow enters through a door in the back of the crate and there must be a bar at the bottom attached to it, 228 mm (9 ins) from the floor and/or 228 mm (9 in) from the back of the door to prevent piglets being crushed by the rump of the sow as she lies down.

The crate permits the sow to get up and lie fully stretched out, but she is unable to turn around. Infra-red heating lamps are usually hung over the space away from the sow to attract the piglets from her side when not sucking. If the piggery is not well insulated or heated, it is advisable for the first few days at least to cover the area where the piglets run with hardboard to conserve heat. The concrete floor, if insulated, can be covered by sawdust or wood shavings, otherwise a covering of straw is warmer.

In the last part of pregnancy sows are very lethargic, but the mammary glands show much enlargement about 24 hours before farrowing begins. Now they show much restlessness and they will go through bed-making movements. By this time the individual teats and the mammary glands have enlarged and stand out. These periods of activity alternate with rest periods, but it is only about an hour before the first piglet appears that the sow will settle on her side and be still.

The length of time that a sow takes in giving birth depends upon her age, fertility, breed and individual characteristics. Gilts, or maiden sows, take longer and often get up and turn over between the birth of the piglets. One large series of observations showed that as many as 45 per cent were born with the hind legs foremost. The sow expels most of her afterbirths along with the piglets, so that the second and third stages of parturition are more or less combined.

Most sows, when they have completed the process of giving birth, will stand and pass a lot of urine before lying again to suckle. It is at this moment that there is likelihood of overlying and crushing some of the last born piglets unless they have got themselves out of the way or been placed outside the farrowing crate by the attendant.

The bitch

The first stage of labour in the bitch reveals itself as increased restlessness; she will not eat, may pant and there is swelling of the mammary glands with milk already present. She will be in her nest for the second stage, the expulsion of the first foetus, which may take up to one hour. The bitch frequently pauses in her straining, once the water bag appears at the vagina, to lick it vigorously, so rupturing it. The head of the puppy then appears and the rest of its body is quickly expelled. Many puppies are normally born in the posterior presentation. There is much variation in the time taken by any particular bitch in giving birth to the rest of her litter. Some may rest for up to several hours between the first and second puppy, others give birth to their puppies at regular short intervals. The bitch expels the foetal membranes within 15 minutes of each birth and they are quickly eaten by her. The dark greenish uterine discharge which may be noted is normal.

Cat

The process is similar to that described for the bitch except that the number of kittens born is often less, rarely more than five. The time taken is less, with more regular time intervals and the uterine discharge is brown.

Care of the Newborn

Whatever the species, the good husbandman will do his best to see that the newborn creature finds its way to the mammary gland of its dam as speedily as possible after its birth in order that it may obtain the valuable colostrum, both for nourishment and to obtain the immunoglobulins that will help to protect it against some of the pathogenic microorganisms of its environment.

Disorders of Reproduction in the Female

Infection Following Parturition

It must be understood that there is a normal and varied bacterial flora in the vagina of all domestic animals. During and after parturition there is inevitable aspiration of some of these microorganisms back into the open cervix and the uterus. The normal defence mechanism against invading microorganisms comes into play and, in the vast majority of instances, within a comparatively short time, i.e. within days, the tract is free of infection.

If there has been difficulty over giving birth, i.e. dystokia, with damage to the epithelium lining the reproductive tract, then a more persistent and serious degree of infection may follow. In cattle particularly, this is likely to be associated with retention of the afterbirth, when a massive infection of the uterus inevitably occurs. The result of this state of affairs is that the dam cannot be mated again until the infection has been overcome and normal oestrous cycling has been established.

The method of treatment varies according to the severity of the case. It is often necessary to remove the retained afterbirth manually if there is a persistent uterine inertia that has not responded to treatment with such hormones as oxytocin and prostaglandin. Whenever, there is evidence of toxaemia in the dam, antibiotics are administered.

Occasionally, with the cow and mare, damage to the vagina and cervix may result in the persistent sucking-in of air, as well as the danger of recurring infection. An operation to repair the reproductive tract becomes necessary.

Another post-parturient abnormality which may occur following the third stage of labour in any species, but is commonest in ruminants, is prolapse or eversion of the uterus. Clearly, it is important to minimize the organ's contamination; so it should be wrapped in a clean towel until veterinary help arrives. In the cow, straining must be avoided during replacement, so a local anaesthetic is injected into the epidural canal at the first intercoccygeal space. After careful washing in warm, normal salt solution, the uterus will be carefully replaced by the veterinarian.

When this accident occurs in the ewe, it is not so difficult to replace the uterus if an assistant raises the hindquarters. If the afterbirth has not separated, the foetal cotyledons should not be forced away from the uterine caruncles, as they will be returned to allow normal separation later.

This is a very rare occurrence in the mare, although it appears likely to be a consequence if birth has to be by Caesarean section because of post-operative suture adhesions.

Infections During Pregnancy

Many infectious agents, bacteria, viruses, protozoa and fungi can prevent pregnancy or induce abortion. They can gain entry to the uterus and there produce a metritis, in two ways: (i) at the time of mating, or from infection introduced during artificial insemination; (ii) from the circulation, brought from a reservoir of infection (e.g. the udder) or from a general blood-borne pathogen.

Mare

There is often difficulty in obtaining a desired pregnancy in the mare, particularly in Thoroughbreds. Most commonly the reasons appear to be of a functional or physiological nature. Venereal infection is generally infrequent. With the exception of the specific venereal pathogen *Haemophilus equigenitalis*, the cause of contagious equine metritis, first introduced into the United Kingdom in 1977, and capsule types 1, 2 and 5 of *Klebsiella aerogenes*, the only other organisms implicated in endometritis in mares are species of streptococci or staphylococci that have managed to gain entrance to the uterus following foaling or as a result of 'windsucking' from defective closure of the lips of the vulva.

The venereal pathogens can gain entrance to the uterus through the stallion carrying infection at mating or from genital examination subsequent to service. To discover the nature of the infection and the appropriate treatment to be undertaken, a sterile swab is carefully introduced into the anterior cervix or uterus and bacteriological examination made. The sinus associated with the clitoris must be swabbed in checking for contagious equine metritis.

Abortion can be caused by infection by the virus, equine herpes virus I or rhinopneumonistis. This produces a mild infection of the lungs and respiratory tract which may not be noticed, but abortion mostly occurs during the last four months of pregnancy. Mares recover spontaneously and after 1-2 months will conceive.

In some countries, *Salmonella abortus equi* still occur although effective vaccines and drugs are available and can bring about its eradication. Again, the protozoon parasite, *Trypanosoma equiperdum*, causing dourine, is still found in some parts of the world. It is a serious and frequently fatal disease and all infected and in-contact animals should be slaughtered.

Cow

Bacteria

There are a number of venereal diseases of cattle, transmission being by the infected male. One of the commonest in both cattle and sheep is *Campylobacter vibrio foetus*, of which three different types are recognized, two found in cattle (*C. foetus venerealis* the most common, and *C. foetus* var. *intestinalis* type I) and the third (*C. foetus* var. *intestinalis* type II) sometimes in cattle but mainly in sheep, where it is not spread venereally. A bull can harbour the infection in his prepuce but show no symptoms. During natural mating the infection is passed to the female and this is likely to prevent conception. If a new bull is used and there is a high proportion of cows and heifers returning to service, this is the infection that should be suspected. Tests can be made on a sample of vaginal mucus and washings from the bull's prepuce using a fluorescent antibody technique. Artificial insemination from a clean bull will permit pregnancy and most cows will overcome the infection and develop an immunity within a few months. Where artificial insemination cannot be used, such as in large beef herds or in developing countries, an effective vaccine is available and can be used to protect cows. Some reports indicate its use in curing bulls of infection.

The commonest cause of abortion in cattle in many countries has been infection with *Brucella abortus*. Brucellosis (Bang's Disease) causes undulant fever in man, although many other symptoms appear apart from a fluctuating high temperature. The cow's udder appears to be the reservoir of infection from one pregnancy to the next.

Infection is the result of grazing over infected pasture or from a cow licking the discharges from an animal that has aborted, or even the foetus itself. If a bull has the infection in his testes, it will have an orchitis and its semen is infective. When heifers or young cows are infected, there is a high incidence of abortions. Older cows that have been infected during earlier pregnancies may carry calves to full term, but these are often of small size and later suffer from scouring. This is a disease which can be passed to humans through the drinking of raw milk or by direct contact by those handling infected cattle.

An effective attenuated live vaccine, S19, has been widely used. It must be injected into calves between 4 and 8 months old. When used conscientiously and consistently this will entirely prevent clinical disease, but does not eliminate the organism. With the greater accuracy of an improved complement fixation test and the use of a fluorescent antibody test, identification of infected cattle has become more certain. This has encouraged the establishment of abortion-free herds. Most temperate countries have, therefore, either already eradicated the disease or have embarked upon long-term eradication campaigns.

Mycoplasma infection (contagious granular vaginitis) is a widespread cause of infertility and occasionally of abortion. It may infect bulls. The organism is *Mycoplasma bovigenitalium* and can be transmitted by the bull, where it causes epididymitis. In cows it is also a cause of mastitis. It is often present in healthy cattle, causing pathogenic changes when resistance is lowered by other causes.

Viruses

Several viruses have been identified as the cause of infertility in cows. A common one appears to be identical with that known to cause infectious bovine rhinotracheitis which frequently appears as a nasal discharge, as well as infectious pustular vulvo-vaginitis (IBR-IPV virus). It is a venereal disease, causing a painful infection of penis and prepuce, making the bull reluctant to serve. When the vagina of a cow becomes infected following mating, the lips of the vulva show swelling and reddening within 24 hours, followed by red vesicles on the mucosa. There is much mucopurulent discharge and sometimes high temperature and lack of appetite. Recovery without treatment occurs within 2-3 weeks and abortion does not normally occur with this form of the disease.

In the U.S.A and some other countries, it appears that a similar virus that causes respiratory diseases can cause abortion following foetal death. If there is delay in expelling the dead foetus, an endometritis may be a sequel. The respiratory form of this disease in bulls can lead to damage within the testes that makes the animal infertile for months.

Another variant of this virus occurs in Africa, causing a specific vaginitis in which there is a copious mucopurulent discharge but without vesicles. It causes epididymitis in bulls and apparently is always transmitted by mating.

There are other pathogenic viruses that have been implicated in causing death of the foetus and subsequent abortion. One is *bovine virus diarrhoea* or mucosal disease, and another epizootic bovine abortion, common in parts of North America and parts of Europe. Two bacterial infections, leptospirosis and salmonellosis that cause serious general signs of disease, will in the pregnant cow bring about the death of the foetus.

Protozoa

A Protozoa parasite, *Trichomonas foetus*, causes another venereal disease, passing from the bull to infect cows and heifers. This prevents conception and, if a pregnant animal becomes infected, will lead to abortion. The disease can be eradicated by employing artificial insemination, being sure to obtain the semen from healthy bulls. If this is done, infected cows will cure themselves; although, if there is a concomitant endometritis with other invading pathogens, the appropriate ancillary treatment must be given. Diagnosis of trichomoniasis is normally by the microscopic identification of the parasite in samples of vaginal mucus or a scraping from the bull's prepuce. Although drugs used to treat human trichomoniasis have been successfully used in bulls, it is generally considered safest to slaughter such animals.

Fungi

Occasionally the spores of a fungus that is found in mouldy hay, *Aspergillus*, are inhaled and, entering the bloodstream, infect the cotyledons of the cow's uterus and can in this manner cause the death of the foetus.

Sheep

There are three common causes of abortion in sheep:

1. Infection with the vibrio, *C. foetus intestinalis*, has already been mentioned, and it appears to be the commonest cause of abortion in intensively managed ewe flocks. Here losses may be as high as 20 per cent, occurring towards the end of pregnancy. Losses in the following year are greatly reduced because of the acquired immunity.
2. Enzootic abortion is caused by an infectious agent that belongs to the *Chlamydia* or PLV (psitticosis lymphogranuloma venereum) group, and on an initial infection of a flock, can cause 25-30 per

cent abortions, again in the last 6 weeks of pregnancy. It is spread by ewes grazing where abortions have occurred or the pasture contaminated by vaginal discharges. It is not carried by the ram. An effective vaccine can be used to prevent or greatly reduce losses in a subsequent years.

3. An organism of the *Salmonella* group, *S. abortus-ovis* may from time to time cause abortions as well as illness in ewes from metritis and septicaemia. The disease is introduced by a bought-in animal who is a carrier. There are likely to be neonatal deaths in infected lambs and persistent diarrhoea causing later deaths with salmonellosis. The organism is sensitive to specific drugs, while a vaccine is available for the protection of healthy young animals.

From time to time, abortions in ewes, still-births and embryonic absorption may be caused by infection with *Toxoplasma gondii*. This is a microorganism that affects many species, including man, and locates in the central nervous system, muscles, the eyes and in the foetal cotyledons. Ewes become immune after infection and if fresh sheep are introduced into an affected flock, this should be done some time before mating to allow immunity to be acquired.

Sow

While there may be certain anatomical or physiological difficulties over reproduction with some sows, likely to be more marked with certain strains within breeds, infertility or abortion from infection is infrequent. In the U.S.A., *Brucella suis* has been widespread, causing inflammation of the testes (orchitis) in boars, abortion and sterility in sows and deaths in newly born piglets. The most effective way of dealing with this disease is to slaughter the whole herd as pig reach marketable weight and delay restocking, after thorough disinfection of the premises, for six months.

There are various virus diseases of pigs, e.g. swine fever, Aujeszky's disease, foot-and-mouth disease, swine influenza, that will prevent conception or interrupt gestation in sows, for they are serious generalized infections.

Bitch and cat

The commonest abnormality affecting reproduction in both the bitch and the cat is a hyperplasia or thickening of the wall of the uterus, often followed by infection, to produce a greatly distended womb. This stage is likely to produce toxaemia and signs of general illness, dullness, lack of appetite and raised temperature. In some instances, there will be a discharge of pus from the vagina.

This condition, especially in cats, is associated with non-breeding, since the effect of repeated cycles of progesterone/oestrogen secretion that occur during the normal cycle of females withheld from mating produces this thickening of the uterine wall with the formation of cysts in a fairly high proportion of older animals. It is generally necessary to perform hysterectomy (removal of the uterus).

Infertility from specific infection is rare, although two organisms are able to cause abortion. The first is confined to canines and is due to *Brucella canis*, infective also from the male, causing orchitis and epididymitis. Not seen in the U.K., it is common in the U.S.A., where infected bitches tend to abort at 45-55 days of pregnancy.

Toxoplasma gondii may occasionally cause abortion, stillbirth and neonatal deaths in both bitches and cats. It can be diagnosed from the multiple small soft nodules on the fresh placenta. Since this organism is transferable to humans, swift action to remove infected animals and careful disinfection of premises must be undertaken.

Other Causes of Infertility

The ability of any species to breed regularly is clearly desirable, but the dependence upon this function is greater in farm animals than in those of the companion animals. If a high-yielding dairy cow does not produce a calf every year, there is a loss of income. The Milk Marketing Board of England and Wales pointed out that in all four common dairy breeds those herds with calving intervals of about 375 days produced approximately the same quantity of milk per cow during the year, no matter in which month they calved. In all breeds, herds that had calving intervals of over 400 days gave less milk. It has been calculated that every day of increase in the calving interval beyond 365 days results in financial loss of 77p, allowing for the costs of feeding. In other words, a single missed oestrus cost the farmer about £16. At this time the average dairy herd in England and Wales contained 44 cows and if its calving index was 395 days, the loss was approximately £1000 per years. Likewise, if a herd of sows does not produce an average of eight viable piglets twice yearly, or the lowland ewe flock a lambing percentage of 150 per cent or better, these farmers' incomes similarly suffer. These aims are more likely to be achieved if the following practical husbandry steps are followed.

Cattle

It has now come to be recognized that the modern dairy cow has a normal conception rate to first insemination of little more than 60 per cent. Those cows that do not conceive to a first service or

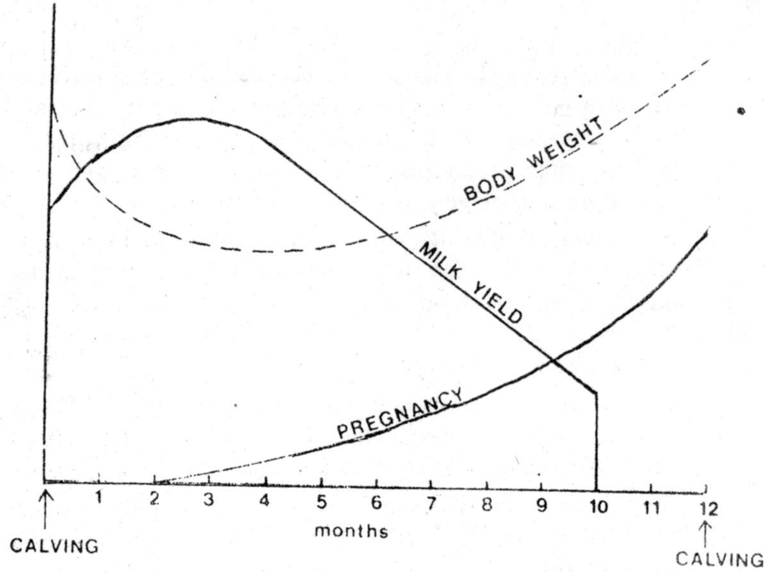

Fig. 6.5. The annual relationship of lactation, pregnancy and body weight.

insemination will likewise have a 60 per cent chance of conception at the end of the next heat period. Except with maiden heifers, dairy cows in North America and European countries are expected to be producing large quantities of milk at the time that efforts will be made to get them into calf again.

Management and conception

There is a continuing management problem because the identification of cows in oestrus is an indispensable step in making sure that the herd's calving index, i.e. the mean calving interval over all the milking cows, is kept down to approximately 365 days. There is often particular difficulty in detecting the heat period of maiden heifers at the age at which the first mating is to be made. Wherever the size of the enterprise merits the keeping of a bull, an effective method is to allow him to run with the heifers. Frequently, a beef bull—the Hereford is a favourite—is chosen, partly because of the smaller calf causing less trouble at birth, partly to obtain a higher price for the calf.

Where artificial insemination is used, there must be careful observation by the stockman to detect the signs of heat. As pro-oestrus ends, cows will show a variety of physical signs, swelling and, in light-coloured breeds, reddening of the vulva. Some cows will have a discharge of clear mucus from the vagina, but most will either attempt to mount other cows or will themselves stand still to permit mounting by other cows. Most oestrous activity occurs at night.

It is of the utmost importance that the stockman regularly observes the behaviour of all cows, and that the date of service for every cow be noted on the daily record sheet that is usually kept in the milking parlour. There are now readily available various designs of wall charts that make it quickly possible to ascertain calving dates and to mark the date at which individual cows must be carefully watched for a return to oestrus.

Electronic aids

Recent rapid increase in the use of microprocessors holds out the prospect that much important information concerning the management and recording of occurrences in each cow, can now be electronically stored and processed with the aid of small computers. Where herds have reached numbers in excess of 200, the cost effectiveness of having correlated information on production, health and reproduction readily available has already been demonstrated.

Functional Infertility

Occasionally cows will be encountered who exhibit abnormal patterns of oestrus as a result of hormonal imbalance. This may take the form of a continuing state of anoestrus—complete absence of heat periods. The cause is likely to be a persistent corpus luteum in the ovary. Most of these cows respond to treatment with the prostaglandin, $F_2\alpha$, or synthetic analogues, e.g. cloprostenol (Estrumate), dinoprost (Lutylase).

Another abnormality is the existence of what is called a cystic ovary. As a result of one or more follicles within the ovary not rupturing in the usual manner at the end of the oestrous cycle, there is a continuation of manifestations of heat. The cow will continue to stand to allow other cows to mount as well as often mounting other cows; the discharge from the vulva persists and there is slackening of the pelvic ligaments and elevation of the tail head. Such a syndrome may be called nymphomania. It is more likely to occur in heavy-milking strains, with cows after they have had their third or fourth calf, and during the early part of lactation as the cow comes up to the peak of her performance.

There is a suggestion that the abnormal quantity of circulating oestrogen responsible for this condition may be due to the high level of prolactin secretion from cows well fed and with high genetic capability for milk production and that this activity in the anterior pituitary may interfere with the secretion or release of sufficient luteinizing hormone to cause ovulation and normal corpus luteum formation, with its production of progesterone. This, in its turn, would inhibit follicle-stimulating hormone.

While treatment at one time was to rupture the cysts manually, with the consequent danger of adhesions developing in a proportion of instances, it is generally believed that injection of luteinizing hormone followed by progesterone is more effective. With the greater availability of luteinizing hormone releasing factor, this alone may be sufficient to get the cow cycling normally.

Embryonic loss

Surveys in many countries have shown that there is, with modern strains of high producing cows of whatever breed, up to a 30 per cent loss of embryos in the first few weeks after mating or insemination. The cause is uncertain but one suggestion is that this is due to the expression of deleterious or lethal recessive genes. With the rapid advancement of cytogenetics and the technique of examining the karyotype (the display of the animal's complement of chromosomes) the incidence of various aberrations has been determined.

Infertility and Nutrition

It has been shown that the energy intake of a high-producing cow during the first 80-100 days after calving, which is the period in which she is steadily increasing her milk output to reach her lactation peak, cannot match the energy being put out in the milk. In other words, she is bound to lose weight. A cow that is losing weight at too fast a rate during this time or is below her normal or optimal weight, is likely either not to show normal oestrus or not to ovulate. She may well stay in anoestrus, that is remain infertile until the lost weight is made up.

This emphasizes the importance of sufficient time, two months being sufficient for most cows, when she should be dry and fed increasing quantities of a diet containing a high proportion of digestible energy, balanced for protein and minerals. After calving, each cow should be fed an amount of a high energy ration calculated to be slightly more than the milk yield expected from each particular cow.

It has been shown that a long-continued deficiency of phosphorus either in absolute terms (less than 0.2 per cent of the diet) or relative to the intake of calcium if the latter is, as occasionally happens, too high, can induce infertility. Such a low effective intake of phosphorus is likely to be associated with diminished intake of digestible energy, since inappetance frequently accompanies this mineral deficiency.

Under exceptional conditions, a continued manganese deficiency may induce infertility. It occurs where there is a high pH of the soil, with excess calcium intake from legumes, which grow well under such conditions, and is accompanied by a deficiency in soil and pasture. Supplementation with the appropriate mineral—dicalcium phosphate, 56 g/h/day for a phosphorus deficiency and manganese sulphate, 4 g/h/day in the second instance—will eliminate these deficiencies.

Production Disease

Two of the three common metabolic disorders that are now frequently referred to as production diseases, viz. ketosis and grass tetany, will interrupt the normal reproduction rhythm. Ketosis is the clinical manifestation of the cow's inability to produce lactose in the udder from blood glucose. This is due to the shortage of oxaloacetate, a necessary substrate in the formation of lactose. This is in short supply because there has been interruption in the sufficiency of supply of the lower volatile fatty acids and glucose coming to the liver from the reticulo-rumen and the intestines. In other words, it is a more acute manifestation of a failure to provide enough digestible energy in the diet. Care in feeding will prevent ketosis, its treatment consisting in providing glucose by injection and a readily fermented material by mouth that will yield glucose, such as glycerol or propylene glycol. Corticosteroids are frequently given to supplement the supply from the adrenal cortex.

Sheep

As with cattle, there is well-documented evidence that there is a moderate level of embryonic death in ewes, probably about 12 per cent. Such ewes may well be successfully served at a succeeding heat period and carry a normal foetus to maturity.

As a result of grazing upon certain strains of subterranean clover in Australia (Dwalganap was a particularly dangerous strain) or of being given a diet containing a high proportion of some tetraploid strains of red clover (Hungaropoly), infertility can be a serious problem. This is due to an oestrogenic substance, genistin, that can lead to cystic degeneration of the endometrium. Avoidance of these strains and the vigilance of plant breeders in the production of newer high-yielding types of legumes, should minimize this danger in the future.

Sows

The farrowing rate as a result of a first service is between 70 and 80 per cent in most intensive pig production units in temperate climates. In the relatively very few enterprises where sows are allowed free-range and boars are run with them, the level of fertility may be expected to be higher.

The economics of pig production encourage the adoption of a system of three-week weaning in order to be more certain of getting two litters annually from each sow. With the older and more orthodox method of weaning at 8 weeks, the sow would show oestrus 4-5 days later, with a higher degree of probability that there would ensue a successful mating. If the system of weaning at three weeks is adopted, two factors of management will help to minimize the time before oestrus is shown. One is to be sure that the sow is fed a 16 per cent protein diet in generous amount and the second is to keep the sow within sound and smell of a boar.

Mare

The fertility of mares, particularly those of the various breeds of ponies, mating under natural conditions, is of the order of 90 per cent. Thoroughbred mares, kept under highly artificial conditions, entailing movement to stud premises for mating, have an average fertility of only 60 per cent. The irregularity of foaling in Thoroughbreds may be due to the infections already discussed, but management factors and a high rate of early embryonic mortality (approximately 19 per cent) play important parts in causing temporary infertility.

The chances of a successful mating are greater if mares are not covered until warmer spring weather and, where possible, the grazing of spring grass has wrought a conditioning effect. Lack of oestrus is the commonest cause of non-infectious infertility in the mare and is often due to lack of condition, i.e. being under weight due to inadequate diet or internal parasites. Bad environmental conditions may play a part too. A proportion of anoestrous mares have been shown to respond to infusion of prostaglandin into the uterus or by the simpler method of intramuscular injection.

Once a filly has had a foal, it is referred to as a 'brood mare' and is likely to be got into foal again as soon as possible. The question of whether a brood mare should be mated at the first or 'foal' heat period which occurs 9-10 days after giving birth to a foal, depends upon several considerations. The most important of these is to be sure that no pathogenic infection has become established in the uterus immediately following foaling, and that the uterus has properly involuted. If the mare is free of infection and is in good condition, she may be teased on the fourth and each successive day after foaling. It is likely that she will conceive at this first post-parturient oestrus.

Bitch and cat

There is remarkably little reliable information about the incidence and the nature of non-infectious infertility in the bitch and cat. Irregularity of oestrus increases with age, bitches over 5 years old having extended heat periods or failing to reproduce altogether. When there is failure of an otherwise normal young bitch or queen to come into oestrus, injections of follicle-stimulating hormone are given. Vaginal smears should be examined and, when the end of pro-oestrus is shown, an injection of luteinizing hormone may permit successful mating.

7

PIGGERY

Domestication of mammals has been an old and profitable business of man. It has been recorded in prehistoric period also. The domesticated mammals, in olden days, were used either for milk, traveling or for meat. There is a long list of animals domesticated by the man. These mammals help the man in various ways. Some of the common mammals which are domesticated by the man are cow, ox, deer, dogs, cats, buffaloes, elephants, horses, donkeys and pigs. Today a number of industries are base on animals and their by-products. Thus these animals are providing handsome jobs to unemployed educated youngmen. Sheeps, goats ans swines are more successful in acilmatising themselves in each and every environment so the industries based on these animals are always in profit and they are preferred by a large number of people. The rearing and development of sheep, goat and swine requires less economies and technology whereas the rate of production is high and duration of maturity is much less as compared with another mammal.

Rearing, development and breeding of swine is known as *piggery*. It is one of the most common and desired industry in developed countries and is gaining momentum in developing countries like India. Pig rearing has never been a new thing for Indian people, but its raising and production has always been in primitive stage. There is no scientific planning and investment of money only because of the reason that it is in the hands of poor and uneducated persons. The higher class of people, educated students and economically sound parties did not show their interest in formation of pig houses and development of a piggery.

The Country Pig

The common Indian pigs are most neglected, scrub animals without any marked characteristic features. They are not paid proper attention and food for their development hence they do not get a fair chance to grow into economical animals and compete with them. These swines are small, black and produce small litters. The meat is also not up to the mark and quite often they are infested with a variety of worms and diseases. They grow slower and die soon. The common Indian pigs have higher fat contents so they are used for medicinal purposes and the flesh is not tasty.

Advantages of pig Production

1. Pig forming requires less money, space and technology.
2. The pigs are developed in comparatively shorter period and with minimum expenditure. They pay the return in the shortest possible time as compared with any other form animals.
3. The housing and equipments are not costly
4. They do not require much labour and attention as compared with any other form animal.
5. The retrun of money is always in a handsome amount as compared with dairy and poultry.
6. The pigs grow faster than any other animal.

Efficiency in Park Production

The profit in production of the pig depends upon the efficiency of the grower in production of the pig and its marketing. He must have a look over the rate of growth, amount of feed eaten and quality of grains used in the feed. The supplementation of vitamins, minerals, antibiotics and proteins in the feed helps in developing the swine faster and such as swine always fetch a good amount of money.

Since the hogs and swines are subjected to a large number of parasites and diseases proper case is to be taken against these diseases.

The swine used in the piggery should be of an improved quality and she must have a good breeding capacity. At the same time marketing of the pig also plays an important role in its economics.

Selection of Breeds

In establishing a piggery the selection of breed plays an important role. It depends on the will of the grower as what type of pig he wants to grow. The profit and loss depends mostly on the selection of the pig. Some of the common breeds aclimatised in India are given below. These breed can be divided into two major groups.

1. The English Class
2. The American Class

The English Class

This class of breed generally includes the pig of English origin. They are well aclimatised in Indian environment and are used for improving the race. Some of the common English breeds are :

(a) *The Large White Yorkshire.* It is an English breed but well established in U.S.A., Canada, India, Ireland and countries of Indian sub-continents. They are used for upgrading the Indian breed. These are white, big in size long and bulky with a long thin and forwardly inclined ears. They measure 350-400 kg.

(b) *Middle White Yorkshire.* A English cross breed boar developed by crossing the white large with Yorkshire extraction. It is a high class bred famous for good quality of meat. It has an early maturing and good recovery. It converts the food into meat in large ratio. These are white, large sized, with a long back and level. They measure about 270 to 380 kg.

Berkshire

This is the oldest breed of swine now established in Australia, Newzeland and Berkshire. It is famous for high yield and is used in upgrading the Indian breed. These can be recognised by its black colour and white patches on the feet, head and tail. The head is short and face is blunt. They weigh 280-300 kg.

American Breed

The American breeds are equally good for upgrading the Indian races and meat production as well. They can be aclimatised easily in Indian environment and are widely used in all parts of India. Some of the common breeds are:

Chester White

A native of Pensylvania U.S.A. is a cross breed of English white Yorkshire with Cheshire and Lincohnshire breeds. These are white in colour and produce big litters. The litters develop fast and reach maturity very soon they measure about 400 kg.

Duroc

Red coloured with white shade in colour. The colour may vary from golden to dark red. They develop fast and produce big litters. They have high milking capacity. These are used in cross breeding weigh about 300 kg.

Hampshire

These are black pigs with white best encircring the body and fornt legs. Head and tail are black, ears straight and erected. Legs are short and produce good quantity of milk and meat as well.

Besides these a variety of porks are found which are equally good in amount of meat and its quality but require more attention. These breeds are Poland, China, Tamworth and Wessex saddleback etc. Day by day improved varieties are being produced which mature early and give flavoured quality of meat in large amount.

Feeding and Management of the Herd

The profit or loss of a pigger is determined on the basis of the total net gain from the piggery. This gain is calculated on the basis of the number of pits marketed. The number of pig produced in the herd depends on a variety of factors and feeding is one of them. Food feeding of the male boar and the sow is solely responsible for producing a big litter. A protein rich diet keeps the male active and smart. It produces a good quantity of sperm with a large number of viable motile sperms, in the same way a healthy female produces a large number of eggs and they are fertilized perfectly. Such a sow produces healthy litter with maximum viability and minimum mortality. Such a sow produces less number of abnormal pigs. The diet of a lactating female is little different with the same given to the male.

The pigs have a very simple digestive system without much modifications and complications, so they require a good and protein rich diet. They are unable to digest more fibrous food but fibrous food up to a limited extent is also necessary for their proper development. A protein rich food helps the female in producing more and more milk and this effects the health of the body sow up to great extent. A healthy baby sow changes the food quickly and starts feeding on grains and fodder where as a ill fed baby sow continues on mother milk for a longer period.

The feed of a sow should have low fibre contents such as cereal grains and their byproducts. It should have proper proportion of minerals, vitamins, medicines, and antibiotics along with the protein. It is to be in a measured quantity, which can keep a pig fast growing, healthy and strong but at the same time should not be much high in fat contents because enough of fat in a sow makes the meat of low quality. There should be a limited ratio of fat and flesh. The balanced diet of a swine comprises of the following ingradients :

Maize, ground barley or grain jowar	=	65-75%
Ground wheat or oat	=	15%
Meat scrap	=	5-7%
Soya meal	=	6-10%
Minerals	=	.5-1%
Salt	=	0.5%

If such a balanced diet is given to a male swine he can be ready and it for service after 8 months whereas a female gets matured in 9 months. It is not advised to use a male in early stage, for better and fruitful results he should be used after one year and that too not daily. He must be given rest at least for 3 days before his use and in the same way the female should not be allowed to male just after her first cycle, she must be spared for 2-3 cycles, i.e., 2-3 months after her first cycle. This cause full development of the uterus and ova and gives better result in days to come. The maturation of male and female depends upon the quality and quantity of the food, they are fed with.

The food of a meat quality of sow varies much from the food of a pregnant or milking sow they require more protein diet as compared with any other normal sow.

Gestation Period

The sow have a heat period of about 40-65 hrs. and the best period of their mating and population is the second half period during the heat period. A pig can easily be recognised during this stage because of her actions. She may mount other swines in the herd and will remain rest less and will not eat. Her appetite is lost. She can be recongnised by examination her vulva which becomes enlarged and congested but this can be done by expert breeders only. The cycle repeats after every third week, otherwise they have gestation period of 114 days on an average which may range from 112-115 days.

Artificial Insemination

Artificial insemination in swines plays a vital role in the development of piggery. In such a condition the boar of a good health and breed is allowed to mate a dummy female with pulsating vagina and the semen is collected in the glass vials. A boar is given at least three days rest after first ejaculation so he can serve only twice a week. The amount of sperms produce varies from 200-250 ml per boar which is kept in freezed condition after diluting with 6% glucose soln. The viability of sperms is 100% for the first 24 hrs after that the viability declines and the chance of fertilizing an egg is reduced to only 50% if such a frozen semen is used after three days. For good litter production a sow is crossed twice after an interval of 8 hrs. This gives good result as the chance of fertilizing an egg is doubled.

In Indian artificial inseminations in swine is not done on large scale because the number of piggeries are not much and if some of them are interested in having a piggery farming they prefer to have a boar also. This ceases the scope of A_1 in sows. Here are some practical problems associated with the artificial insemination of the swine such s the sperms have high mortality and poor ability of frozen sperms. The sow have small vagina which does not exhibit the visual symptoms of heat and they too show poor heat detections.

The sows in a herd have different days of heat so they can be managed by the boar.

The scientists in England have developed a new chemical which is fed to the female swine regularly and till the date it is fed to the sow she does not come into heat and the day the dose is checked the sow comes in heat. This has become a boon for piggery farming. The breeders check the heat period chemically and allow all the females of the herd to come in heat on same day. This helps in A_1 and all the females give birth to the hogs on the calculated day. This provides the breeder with a benefit of arranging and managing the females of same category. A particular type of bed. Feed, heat, water, mineral, vitamins and aids is given to the female during her pregnancy. This saves a breeder from arranging various types of beds, feeds, aids and at the same time large number of baby pigs are obtained which are marketed in a lot and the profit is obtained in a handsome amount. These things are still needed in India to improve the piggery farming.

The baby pig is given a good nutritive diet to develop fast and attain good size. Fatty diet is given only at certain intervals and after recommendation of the doctors. The well developed hog is now send to slaughter house or market. The slaughter of a pig is done after starving the pig for 24 hrs. This gives a good and standard product.

Growth of Young

For the good growth of young ones, supplementary feed should be given after 25 day by creep feeding. It should be kept in mind that sows fed on synthetic diet may not be allowed to become much fatty otherwise the size of the litter would be smaller. The good growth in the early developmental stages plays very significant role for the development in length, muscle, weight of the pits, so up to the age of 25 days young pigs should be fed on highly nutritive and easily digestible diet. The young pigs, with the weight of 50 pounds feeding on less nutritive diet, have good ability of fast growth to attain the maximum weight provided a good nutritive diet is given to them.

Proportional Growth of the Body

The proportional growth of the body of pigs varies with the age i.e.,the head and the legs are larger in comparison to the body but the age, the body grows very fast as a result the muscles of thigh become concave. Such growth is found more in middle white pig in comparison to large white pig. It is very useful to have an idea about the change in the weight of the developing pigs and a proper control on that change.

The growth in pigs runs from the cranium towards backwards and from the tail forwards which meet in the lumber vertebrae as a result the loin region grows very fast followed by the pelvis and the thorax than the neck. The head grows very slowly, it is advisable that the pigs in early stages should be allowed to fed on poor diet due to which early developing parts, as head and legs would grow very slowly and if later given highly nutritive diet the parts which grow later i.e., loin would grow very fast and the carcass with high proportion of fat is produced. In the young pigs the fat is soft and in the older ones it becomes firmer when growth rate is very fast and fat is synthesized from the carbohydıates.

Slaughter of Pigs

Prior to the killing, the pigs should be kept on starvation for 24 hours and given complete rest to get standard product. The slaughtering of pig is carried out by making it unconscious by a hard stroke given on the head, and piercing a double edged knife in the neck to shed the blood out of its jugular veins. Now carcass is washed, cleaned well in hot water and cut open to separate the various parts for different preparations.

Products of Piggery

Pork

The flesh of pig is known as pork in general and the fleshes obtained from different parts of the body have been given different name e.g., Bacon obtained from the back and sides and Ham from the back of the thigh. According to the data produced by the Directorate of Marketing and inspection, Government of India the production of pork is 5 percent of total production of meat in India.

Bristles

The wiry and stiff haris of the pigs, hogs and wild boars are known as bristles and they are obtained from the back and neck. Short bristles are obtained from the flanks and belly of the animal. The bristles obtained from the living pigs are superior in quality than those of slaughtered ones. The rough and coars bristles are generally used for varnish work and painting brushes.

The main bristle producing areas are Uttar Pradesh, Madhya Pradesh, and Punjab. The principal trading centres are Kanpur and Jabalpur but 70 percent of the total bristle export is being made from Kanpur.

Two types of bristles viz., *Desi bristles* obtained from pigs of Uttar Pradesh and Madhya Pradesh, and *Darjeeling bristles*, obtained from Himalayan foot hill and Darjeeling district. The bristles of three colours are usually found viz., white, grey and black.

The important bristle markets in the country are at Agra, Allahabad, Azamgarh, Faizabad, Jaunpur, Amaraoti, Nagpur, Santhal Parganas, Caluctta and Kakinada.

Mumbai is the biggest port to export the bristles worth millions of dollars to U.S.A., U.K., Germany and Japan. Indian Standard Institute, New Delhi standardizes the bristles and provides *agmark*. Only the properly graded bristles under 'Bristle Grading and Marketing (amendment) Rules-1962' are permitted for export purposes.

Sausages

Sausages are prepared by fresh minced pork, free from bone and skin. For the preparation of sausages usually shoulder piece, ham and bacon are preferred. The sausages are prepared by washing, cooking and mixing the pork with spices such as white pepper and paprika to make it delicious.

Lard

The fat of pig, squeezed from the body tissue is termed as lard. For the preparation of lard, carcasses are cut into pieces, minced, boiled over a furnace and the fat is removed-carefully when it is boiling hot.

Lard is used as a fine cooking medium and in the manufacture of soaps, lubricants, greases, candles and water proof materials.

Other Uses

The most important pharmaceuticals derived from thyroid, pituitary and pancereas of pigs are pepsin, thyroxin, pitutrin, insulin, liver extract, testesteron etc. The pig toe-nails are used for making tobacco fertilizer, plaster and other plastic materials. The blood is used as food for livestock and poultry and also as a manure.

Diseases and Control

The swine production suffer heavy loss due to diseases and parasites. These parasite kill the baby pigs and check proper growth and development of the pig as adult. They damage about 60-65% of the total pigs produce, either directly or indirectly. Their control is a major factor in swine production.

Most of the diseases can be controlled by applying preventive measures and proper sanitation. Over crowding, mismanagement, roaming of pigs in open places should be checked. Vaccination and medication at regular intervals helps in preventing the diseases.

In India swines are subjected to various diseases like, plague, flue, fever, pox, foot and mouth diseases, cholera, navel ill and dysentery and tuberculosis etc. These diseases are contageous diseases and can be checked by proper sanitation and care.

The infected pig becomes weak, looses its weight, the loss in appetite and vigour is well marked. He may sit or stand alone in an unusual manner. The colour of urine and faecal matter changes and some times it is supplemented with mucous and blood also. The colour may be red, green or black. Coughing and eye and nasal discharge is well marked in cold and flue cases. They stop feeding and go for water quite often. Parasites also play a vital role in swine business. The common parasites are round worm, hook worm, type worms, lungworm and mange. Ticks and mites, flea, mosquitoes, lice and bugs are common ectoparasites on pigs which may cause skin disease and itch.

These worms and ecto and endo parasites are reduced in number if proper care is taken in the management of a pig house, ventilation, sanitation and space have their role in minimising the parasites. These houses should be cleaned 2 sprayed with DDT, BHC or Gamexene.

8

BUFFALO

The domestic buffalo or water buffalo belongs to the *Bovidae* family and from time immemorial has been a domestic animal in India, Malaya and Egypt. It occupies an important place amongst the domestic animals of the tropics as a provider of dairy produce, beef and draught power. Its greatest asset as a domestic animal is its ability to subsist on the coarsest fodder and to convert it most efficiently into animal produce.

The water buffalo is found in a more or less domesticated state throughout the northern tropical and sub-tropical zones in Asia as well as in the Philippines and Trinidad and in all countries of the Mediterranean basin except France. In the southern tropical zone it is found in Indonesia and it has been introduced to Melville Island near the northern coast of Australia. It is also found wild in the Northern Territories of Australia but the original animals were domesticated.

In recent years, the buffalo commonly known as an 'Asian Animal' has attracted global concern. The buffalo is the dairy, draught and meat animal of Asia. The Indian subcontinent is the home tract of the world's dairy buffaloes. To-day in India, the water buffalo is recognized as her milk machine. It accounts for more than half of India's total milk production, although it constitutes only one third of the total milch population. Nevertheless, its potential for milk production remains only partially tapped, as almost 80 to 85 per cent buffaloes are non-descript, yielding on an average no more than 1.5 litres of milk per day in contrast to an average of 7 to 8 litres of daily milk yield obtained from well bred milch breeds.

The domesticated buffaloes may be classified into two main categories, viz., (1) the swamp buffalo (chromosomes, 2n = 48) and

(2) the river buffalo (chromosomes, 2n = 50). They belong to the same species, *Bubalus bubalis* but have very different habits.

Some Common Terms in Relation to Buffalo

Details	*Expressions*
Species called as	Bovine of Butaline
Group of animals	Herd
Adult male	Buffalo bull
Adult female	She-buffalo or buffalo cow
Young male	Buffalo bull calf
Young female	Buffalo heifer calf
New-born one	Buffalo calf
Castrated male	Buffalo Bullock
Castrated female	Spayed
Female with its offspring	Calf at foot
Act of parturition	Calving
Act of mating	Serving
Sound produced	Bellowing
Pregnancy	Gestation

As the name implies, the water buffalo, whether of the river or the swamp type, has an inherent predilection for water and loves wallowing in water or mud pools. The river buffaloes, as a rule, shows preference for clean running water, whereas the swamp buffalo likes to wallow in mudholes, swamps and stagnant pools. Like cattle buffaloes are good swimmer.

The river and swamp buffaloes vary slightly in body structure, size and colour. A swamp buffalo is recognised by its short stocky body, short face with wide muzzle and short thin legs. These buffaloes are usually dark grey in colour, although variations from black to albinoids are found among the population. Swamp buffaloes vary in size from 300 to 600 kilograms. They are primarily used as draught animals, but they do provide small quantities of milk for a farmer's family.

Breeds of Indian Buffaloes

Only in India are there found well-defined tropical breeds with standard qualities and, even there, well-bred individuals are far out numbered by nondescript animals. In other countries, although distinct types are to be found, no type has the specific physical characters which mark it from other types so that it can be unquestionably recognized.

Good, well-defined, milk breeds are found only in the Punjab, Rajasthan, and the Gujarat district of Bombay, but in Central and South India there is a noted draught breed.

These animals belong to a non-descript class and these vary greatly in size, weight and general features known as *deshi*. However, on the basis of regions the well defined buffalo breeds are as follows:

Table 8.1

Murrah Group	*Gujarat Group*	*Uttar Pradesh Group*	*Central India Group*	*South India Group*
Murrah	Surti	Bhadawari	Nagpuri	Toda
Nili-Ravi	Jaffarbadi	Tarai	Pandhepuri	South Kanara
Kundi	Mehsana		Manda	
Godavari			Jerangi	
			Kalhandi	
			Sambalpur	

Murrah

The *Murrah* is the most important Indian breed of buffalo, and is the most efficient producer of milk, not only in India but probably in the world. Bulls of this breed are used extensively for up-grading inferior stock. She-buffaloes are used in most of the important cities for the supply of milk and ghee.

The home of the breed is mainly in Punjab and Delhi, but animals are bred pure in the United Provinces, Rajasthan and other places. Rohtak in Punjab is a well-known market from where thousands of high yielders are exported.

The Murrah has a deep massive frame with short, broad back and a comparatively light neck and head. It has short, characteristics tightly curled horns, well-developed udder and long tail with a white switch reaching to the fetlock. Other features are: short massive limbs with good bone, broad hoofs and drooping quarters. The colour is usually black, but fawn-grey colouring of the hair and white markings on the face, legs or tail are not uncommon. 'Wall' eyes are also not unusual.

The skin is soft, smooth with scanty hair. The body weight of bulls amounts on the average to 550 kg that of the she-buffaloes to 450 kg. The height at withers on the average is 1.42 m for bulls and 1.32 m for she-buffaloes.

Production

The average milking capacity in established herds is about 1400 to 2000 kg with a butter fat content of 7 per cent of milk in a lactation of 9 to 10 months.

Nili/Ravi

The *Nili* and *Ravi* are two types of buffaloes found in the valley of the rivers Sutlej and Ravi in the Districts of Montgomery and Ferozepur. There is no essential difference between the two types although for a long time they were treated as different breeds, but closer study indicated that it was a distinction without a difference, and now they are officially treated as one breed.

This is considered to be one of the best breeds of buffaloes in India, next only to the Murrah. Like the Murrah large numbers are exported annually from the breeding areas for milk production in cities. The average milk yield is practically the same as that of the Murrah.

These animals have a medium-sized, deep frame with an elongated, coarse and heavy head, bulging at the top, depressed between the eyes and ending in a fine muzzle. Horns are small with a high coil and the neck is long, thin and fine. It is the face and forehead that distinguish this breed mainly from the Murrah.

The typical Nili/Ravi cow has a well-developed udder. The tail is long, almost touching the ground.

The colour is usually black, but brown is not uncommon. Pink markings are sometimes seen on the udder and brisket. White markings on the forehead, face, muzzle and legs, white switch and wall, eyes, which are also seen, are much liked by breeders. The average live-weight and measurements of mature animals are as follows:

	Male	Female
Live-weight : pounds (kg.)	1,300 (589.7)	1,000 (453.6)
Height at withers : inches (cm.)	54.0 (137.2)	53.5 (135.9)
Length from point of shoulder to pin Bone : inches (cm.)	62.5 (158.8)	58.5 (148.6)

Surti

The *Surti* is a popular breed found in the 'Charottar' tract of Gujarat in Mumbai State between Mahi and Sabarmati rivers. The

Surti is known to be an economic milk producer and the average lactation yield is reported to be 3,650 pounds (1,655.5 kg.) milk with 7.5 per cent fat.

Surti buffaloes are well-shaped animals of medium size. They are rather low on the legs and have a mild and placid disposition, but the general appearance is bright, with prominent, round, rather bulging eyes. The horns are of medium length and are sickle-shaped. The colour is black or brown; good specimens have two white collars, one just round the jowl from ear to ear and a second lower down around the brisket akin to the breast plate worn by horses. No other breed of buffalo has as straight a back as this one. The udder is well shaped and well developed with pinkish-coloured skin and medium-sized teats, placed squarely. The skin is fairly thick but pliable, soft and smooth, with scanty hair.

The average measurements of mature animals are as follows:

	Male	*Female*
Height at withers : inches (cm.)	51.5 (130.8)	49.0 (124.5)
Length from point of shoulder to pin Bone: inches (cm.)	56.3 (154.2)	54.5 (138.4)

Jaffarabadi

Jaffarabadi buffaloes are very massive animals and are seen in their purest from in the Gir forest of Kathiawar where they are bred in large numbers, almost solely for ghee production. They consume large quantities of fodder but the milk yield is high and the butterfat content exceptionally high. Their noticeable feature is the very prominent forehead and heavy horns which are inclined to droop on each side of the neck and then turn up at the points but not in such a tight curl as in Murrah buffaloes. They are usually black in colour.

Mehsana

This breed is found in districts of Bombay State and neighbouring areas. It is considered to be an intermediate type between Surti and Murrah buffaloes but here is a considerable amount of variation from one district to another. The horns of these animals are usually curled after the manner of Murrah buffaloes but not so tightly. They are either black or fawn-grey in colour with usually some white marking on the face, legs, or tip of the tail. The udder is usually well shaped with uniformly placed teats. Mehsana buffaloes are considered to be particularly valuable milch cattle because of their early maturity, persistence in milk production and regularity in breeding. They are of medium size, economical to feed, and are used largely for milk and ghee production in Bombay state.

UTTAR PRADESH GROUP

Bhadawari

Bhadawari estate of Agra district and adjoining areas of Gwalior and Etawah. Animals are found scattered in the surroundings of Jamuna and Chambal rivers.

Medium size and wedge-shaped body. Comparatively small head bulging towards horms. Legs are short but stout. Hooves of the hind quarters are more backward than fore quarters. Colour is copper, hair scanty. Barrel is short but well developed.

Table 8.2

Trait	*Range of vale*
Age at first calving (months)	48.3—50.78
Lactation yield (lit.)	1111—1252
Lactation length (days)	276
Calving interval (days)	453.6
Dry period	156

Forehead is slightly broad and deep in the middle. Face is comparatively narrow with slightly marked nose. Eyes are prominent, active and bright. Udder is not so well developed but milk veins are fairly prominent. Teats are of medium size and uniform in length.

Average yield ranges from 2,000 to 2,070 kg. in a lactation period of 305 days with a high fat percentage. Males are used for draught purposes.

Tarai

The breed gets the name from the *Tarai* area of U.P., where it is mostly found between Tanakpur and Ramnagar. The breed is native to hilly area. They are frequently crossed with Murrah bulls.

It has a moderate body, slightly convex head with prominent nasal bones. Horns are long and flat with coils, bending backwards and upwards having pointed tips. The eyes are rather small but ears are long and coarse. Legs are short but strong. The tail is long, reaching below the hocks. The colour of the skin varies from black to brown. Sometimes there is a white blaze on the forehead. The switch of the tail is white.

The breed is poor regarding milk production, which may be as low as 2-3 kg daily. Males are efficient draught animals used for agricultural operations as powers including transport. Tarai breed is well adapted to the difficult environmental situations and nutritional conditions of the Tarai.

Central India Group

Nagpuri

Synonym, Marathwada, Berari, Ellichpuri, Gaulani, Varad, Gauli.

As the name applies, these animals are mostly found in Nagpur including Wardha, Yeotmal, Akola, Buldana, Amrawati and Achalpur in Maharastra. The breed has got several strains differing in colour and other superficial traits.

The colour is usually black; occasionally white markings are observed on the face, legs and switch. In general the breed is known as lighter type than those of north, not so squat and have a short tail. The face is long and thin with a straight profile. The neck is also longer with heavy brisket. It has long horns flat-curved reaches towards back over the shoulders. Naval flap is short or almost absent.

Male body weight averages 525 kg. while female attains about 425 kg. The average height at withers in male 142 cm and that of cows 132 cm; heart girth is 210 and 205 cm. for male and female respectively.

The females are moderately good yielders, producing on an average 1,000 litres milk per lactation of 300 days with butter fat content between 7.0-8.5 per cent. The males are used for heavy draught but are slow.

Pandhepuri

Synonym Pandharpur, Dharwari

The breed is widely distributed in south Maharastra, parts of Andhra Pradesh and Karnataka States.

Animals are of medium size with long narrow face and very long, flat and usually twisted thin horns.

Males are hardy and well suited for drought purpose.

Manda

Synonym, Parlakimedi, Ganjam.

The animals are bred in the hills above Parlakimedi and Mandasa on the borders of Orissa and Andhra Pradesh. Basically the breed is reared in areas of thick forest usually on natural herbage and brought down to the plains for sale.

The general colour is brown or grey with yellowish tufts of hair on the knees and fetlocks, and the switch is yellowish white. The breed is a medium sized animal. The eyes are sharp with a broad red margin around the lids. The horns are broad and semi-circular extending backward and inward. The forehead is flat with short muzzle. Neck and forelegs are also short but stout with well developed chest.

Milk yield is satisfactory. Males are hardy and like a bullock can work in the hot sun. It can draw a load of about a ton but at a slow pace.

Sambalpur

Synonyms, Kimedi, Gowdoo.

The home tract of this breed is controversial. Originally it was known to be habitat of Sambalpur area of Orissa, later on it has been suggested that the main habitat of this breed is around Bilaspur district of Madhya Pradesh wherefrom calves are brought by 'gowdoo' (Herdman) to Sambalpur area.

Animals are large and powerful having long, narrow barrel and prominent fore-head. Body and coat colour is generally black but it varies to brown and ash grey.

Males are very active and good for drought purposes which are affected by high atmosphere temperature. Females breed regularly and produce milk satisfactorily.

Kalhandi

Synonym, Peddakimedi.

The home tract of this breed has been referred to the eastern part of Andhra Pradesh and adjoining areas of Orissa.

A strong with broad horns and half curved running backward at the tip. Body colour is between grey and ash grey. It has well developed chest and strong fore-limbs. Eyes are prominent and large with narrow red margin around the lids. Tail is of medium length with white switch. Due to light colour it tolerates heat more than dark coloured buffaloes.

Animals are docile and hardy. They are used to carry loads in hilly areas as well as for ploughing in plains. Milk yield is satisfactory.

Jerangi

This breed of buffalo is widely distributed in Jerangi hills of Orissa and northern parts of Visakhapatnam and west of Ganjam in Andhra Pradesh. One of the dwarf breeds of buffalo and its height does not exceed four feet. Horns are conical and small, and run backward, body colour is black.

Not that much good in milk production but are useful animals for ploughing in water-logged paddy fields.

South India Group

Toda

The name of the breed originates from the name of *Toda* tribes of Nilgiri hills of Tamil Nadu State who rear this breed.

These are large-sized animals having long barrel and strong build. Horns of these buffaloes are variable in shape but are usually set wide apart run outward, upward and then inward at the top. Face is short and wide. Hump and dewlap are absent, chest is broad and deep; legs are short and sturdy. Along the crest of the neck, hump area and back, there is a thick growth of hair like a mane which imparts a bison like appearance.

The animals are not always docile. All of sudden they may become furious and dangerous particularly to unknown persons. When necessary, they fight against even tigers and other wild animals in an organised team.

Females are good milkers, yields between 4.5-8.5 litres daily with an average of 7.0% fat and possesses typical well-flavoured taste.

Breeding and Management Problems

Buffaloes are Seasonal Breeders

In tropical countries like India, *she* buffaloes indicate seasonal sex periodicity in respect of oestrous and conception. As much as 62.8% animals show periodicity in October to February with the peak around December and this period coincides with higher conception rate. During the period from March to September, with comparatively longer days and higher intensity of sun, ovarian activity appears to be adversely affected.

Buffalo bulls are sexually least active during hot seasons. High environmental temperature upsets the normal physiological functions affecting spermatogenesis in males and ovarian activity in the females. 82% of services take place in November. Even under A.I. conditions, 65% of the inseminations of the year are being carried out during the months of September to February. Thus 80% of buffalo calvings are recorded only in the months of July to December and minimum in hotter months. With judicious management practices such as heat detection, protection from thermal stress, adequate and balanced ration and optimum time of insemination, the calvings in buffaloes could be evenly spread round the year to a great extent.

Silent Heat

Silent heat refers to normal follicular development and ovulations without the behavioural signs of estrus.

In summer the extreme climatic stress coupled with increased day light reduces the incidences of buffaloes coming in regular heat. During this time other reproductive problems like repeat breeding and pregnancy losses are also increased. It is generally felt that most of the buffaloes do have sub-functional ovaries causing weak or silent oestrus (in 60% cases) and such incidences are even higher than the true anestrous, especially in summer.

The research workers are trying to find out the causes of these disorders of ovarian origin and are successfully trying to correlate them with the lack of management, nutritional status of the animal and hormonal (serum gonadotrophins and steroids) profiles.

However, if due to excessive environmental stress the CL in buffaloes does not regress or partially regresses (in non-pregnant condition), it will continue to secrete progesterone and will inhibit further release of LH, thus the buffaloes will go in a state of anestrus or subestrus conditions, so long as CL does not completely regress.

Corrective Measures

For animals which either do not manifest oestrus or are in a state of silent oestrus or sub-oestrus firstly need managemental attention rather than therapies.

Two or three examinations with an interval of a week between each, will reveal that most of them are cases of apparent anestrus. In such cases simple utero-ovarian massage and close detection of oestrus will elleviate these conditions in most of the cases.

In extreme summer condition these animals should be protected from direct solar radiation by maintaining them in shaded half walled sheds. If possible animals be given showers in addition to wallowing facilities particularly at mid-day during April, May and June.

Parading vasectomized bulls for heat detection can be gainfully employed at organised/large farms. Since in summer buffalo bulls soon lose libido, replacement by young high libido bulls may be thought of where it is applicable.

Painting of Lugol's Iodine

The external is swabbed thoroughly with Lugol's solution (B. Vet. Co, Iodum 5 g, pot, iodide 10 g, distilled water 100 ml) with the help of a metallic swab holder (46 cm.) and vaginal speculum- double application of this solution is sometimes recommended at week's interval. It is presumed that application of Lugol's Iodine on the cervix causes local irritation and bring about reflex stimulation of anterior pituitary for secretion of gonadotrophins and thus cyclicity starts.

Some workers have used estrogen and progesterone in anaestrus and sub estrus cows with encouraging success. The idea behind using small doses of these hormones are to act as a positive fed back at the hypothalamus level to release gonadotrophin releasing hormone (Gn-RH) and simultaneously sensitize the gonadotrophas in the anterior pituitary so that Gn-RH may act fully to stimulate more release of gonadotrophins.

It should be noted that the method can never be considered as a substitute for good management since use of hormonal therapy might also lead to inherent weakness in the progenies.

Use of prostaglandin ($PGF_2\alpha$) induces luteolysis resulting in sharp fall of serum progesterone level and thus helps to bring back normal oestrus cycle. Like hormonal therapy, prostaglandin therapy also can not be recommended as a substitute of good management.

Repeat Breeding

A repeat breeding in cow or buffalo is the one which has normal oestrus, oestrus cycles as well as reproductive tract and has been bred three or more times by a fertile bull or semen yet failed to conceive.

It is possible that pathology which cannot be detected by normal clinical methods may be present which include among others, failure of fertilization, an ovulation and delayed ovulation, tubal obstructions, early or latent embryonic mortalities, poor breeding and management techniques including genetic, nutritional and infectious factors. Endometritis in many cases has been found to play major factor for causing repeat-breeding. Endometritis is caused by sporadic uterine infections by a varieties of microorganisms such as *C. pyogenes*. *Coliforms*, *Pseudomonas aeruginosa*, *Streptococci* etc., viral and fungal agents along with mycoplasmal agents may also cause endometritis in cattle and buffaloes. The condition follows mainly parturition, especially abnormal ones; abortion dystokia, retained placenta, genital prolapse, uterine inertia, traumatic lesions in the uterus, cervix and vagina. Endometritis may also occur in cows and buffaloes after coilus or after AI under unhygienic conditions.

Appropriate antibiotic treatment is advocated after conducting antibiotic sensitivity tests where repeat breeding is associated with infection of the tubular genitalia.

Daily examination of repeaters is necessary to correct the conditions arising from ovarian dysfunction. Anovular heats can be detected from the absence of palpable C.L. during dioestrus. Intramascular injection of 25 mg of progesterone during early heat may help inducing ovulation.

The golden rule, prevention is better than cure applies more appropriately to reproductive problems. The incidence can be kept under control by adopting methods such as: (1) improved feeding and management, (2) providing hygienic surroundings at the time of calving, (3) asceptic precautions at the time of inseminations, and (4) educating farmers on detection of heat and maintenance of breeding records.

Weaning

Separating calf from its mother at a very early age is a bit difficult in buffaloes due to high motherly instinct. Since the method is important to assess dam's milk production and to feed calf according to live weight a substitution technique could be followed.

A single calf is substituted for let-down of milks in a number of buffaloes and the real calves are weaned just after birth. It can be achieved by putting blinkers (materials used to shut off the side views of buffaloes) during parturition and subsequently soiliing the body of substituted calf with placental secretions. The dam starts licking the substituted calf after removal of blinkers.

Calf Mortality

Buffalo calf mortality rate under one month of age averages about 10% and varies from 3-30% in individual herds. Losses upto 50% have occurred in large dairy herds. A calf mortality rate of 20% can reduce net profit by 38%.

A number of factors considered responsible for this hazard, are discussed below:

Antibodies are not transferable from buffalo dam to her fetus through placental membranes and thus they are susceptible to various virulent diseases. In buffalo calves inspite of feeding colostrum, a low antibody titre exists. Immunoglobulin levels have been reported to be 29.73 mg/ml it day olds and the quantity increase to 35.66 mg/ml on the second day of life. On the other hand in one estimate it has been noted that colostrum which is the only source of immunoglobulin for the buffalo calf, contains 68.75 mg of immuglobulin per ml on the first day, 23.75 mg/ml on the second day and 1.01 mg/ml on the fifth day of lactation.

Certain meteorological influences may have an effect on calf mortality rate. During the winter months, mortality may be associated with the effects of cold, wet and windy weather while during the summer it may be the hot, dry weather.

Most of the deaths occur during autumn and winter months before the age of 3 months. The causes of mortality in order of priority have been found (i) Pneumonia, (ii) Enteritis, (iii) Toxaemia/Septicemia, (iv) Worm infestation, (v) Bloat etc.

The cause of high mortality in male calves could also be due to neglecting tendency by the management particularly regarding feeding.

Some of the prescribed managemental practices including feeding practices may be followed for minimizing the mortality rate.

Housing

In commercial herds, calves after weaning may be kept in groups in large pens where individual feeding is advocated. In small herds calves after weaning should be kept in separate pens (24 sq. ft.) upto 3 months of age to avoid suckling instinct. The methods will eliminate the possible calf scour and parasitic infestation.

Feeding Colostrum

It is the milk secreted by the udder immediately after parturition and for the following 3-5 days. It contains 20% or more protein a little more fat, 10 to 100 times more vitamin A. three times more of vitamin D and may be tinged pink due to blood corpuscles. It acts as a natural purgative for the calf, cleaning from its intestines the accumulated faecal matter (meconium). Of much greater importance, it is through the medium of the colostrum, antibodies which protects against various bacteria and viruses are supplied to the new born calves. Colostrum coagulates at about 80^0 to 85^0 C and can not therefore be boiled.

The calves at birth after weaning should be fed colostrum within 2 hrs. and its feeding should continue for a period of five days at the rate of 1 to 1.5 kg per day.

In case the dam does not give colostrum a substitute of equal nutritive value prepared from 2 eggs and an ounce of castor oil may be fed for builiding up resistance. It may be necessary to inject dam's serum to such a calf for augmenting antibody titre in the body. The feed efficiency ratio of buffalo calves is as high as 1 kg. gain per 1.16 kg. dry matter. Milk feeding schedule is given in tabular form. It is necessary to provide milk to weaned calves at body temperature preferably with supplements to make up the deficiencies of Fe, Cu, Mg, Zn and Mn. Green fodder containing upto 100 gram dry matter may be offered daily from 15 days of age onwards so that it stimulates early rumen development.

Table 8.3. Milk feeding schedule

Age (days)	*Colostrum (kg.)*	*Body weight (kg.)*	*Milk (liters)*
1–5	1 to 1.5	—	—
6–15		Upto 25 kg.	1.0
11–15		26–30	2.0
15 and above		31–40	2.5
		41–45	3.0
		46–50	3.5
		51–55	4.0
		56–60	4.5
		60 and above	5.0

Feeding Antibiotics

Antibiotics are necessary in calves below 3 months of age to overcome certain stress conditions developed due to clinical or sub-clinical type of scour. These drugs are usually unnecessary in calves above 3 months as by this time rumen starts functioning and antibiotics are liable to interfere in normal functioning of ruminal microbes.

Feeding Milk Replacer

The objectives of feeding milk replace to calves are primarily to reduce the cost of raising buffalo calves and also to save milk for human consumption. Milk replaced may be fed as detailed in Tabular form. It may be noted that there is a gradual decrease in milk and corresponding increase in milk replaced at the rate of 200 g milk substitute per kg milk reduction. The general principle to be followed for calculation of milk requirement is that *it should be half kg. less than one tenth of the body weight*. It is also necessary to give a minimum of 1 litre of milk per day with milk replaced and the mixture should be diluted about six times with warm water to provide a drink in suspension form.

DISEASE

The most prevalent diseases of buffaloes are much the same as those of cattle, viz. rinderpest, foot-and-mouth disease, haemorrhagic speticaemia and anthrax. Buffaloes are generally less susceptible than cattle to foot-and-mouth disease but, when contracted, its debilitating effect may be disastrous: an attack during growth may cripple the animal for life, and if it occurs soon after calving often the entire lactation is affected. Buffaloes are much less susceptible than cattle to rinderpest, when managed under comparable conditions; the Egyptian buffalo is reported to be immune to natural infection. They are often infects with trypanosomes but the disease runs a chronic course giving rise to no visible symptoms. They are also infested with most of the larger parasites of cattle but the extent of the damage caused is often less noticeable. As would be expected, the incidence of liver fluke is higher than in cattle. Of the arthropod parasites, the stable fly (*Stomoxys calcitrans*) and the so-called buffalo fly (*Siphona exigua*) attack buffalo more than cattle. On the other hand buffaloes are less affected by ticks but, as calves, they may be sorely tried by lice and mange mites.

9

HORSES AND MULES

The proper evaluation and selection of either purebred or grade draft animals is a much more complicated job than that of properly evaluating other farm animals. In addition to the conformation, quality, and condition of the horse or mule, the buyer or judge must also consider the feet, legs, action, wind, and temperament. The importance of these last items is reflected in the old sayings "No foot—no horse" and "The feet, legs, and action are half the horse."

To properly evaluate draft animals one must first learn the proper names for the different parts, and then he must consider the purpose of work stock and how they accomplish their purpose. Draft horses and mules are designed principally to pull more or less heavy loads at a medium to slow gait. Practically all of the power for this pull is generated in the rear quarters of the horse and mule.

The pull is exerted almost entirely through the rear feet with the front feet serving mostly to support the front part of the animal as would a set of wheels. From the rear feet the force is transmitted through the rear legs, loin, back, and shoulders into the collar. These facts warrant the drawing of several conclusions:

1. The feet of a draft animal must not only be sound but should also be large so that they can obtain maximum grip on the ground.
2. The bone of the draft animal must be strong and durable and the legs must be placed or "set" properly.
3. The hocks must be broad, deep, and clean because these are the levers through which the power, generated in the muscles of the rear quarters, is transmitted to the feet when pulling.

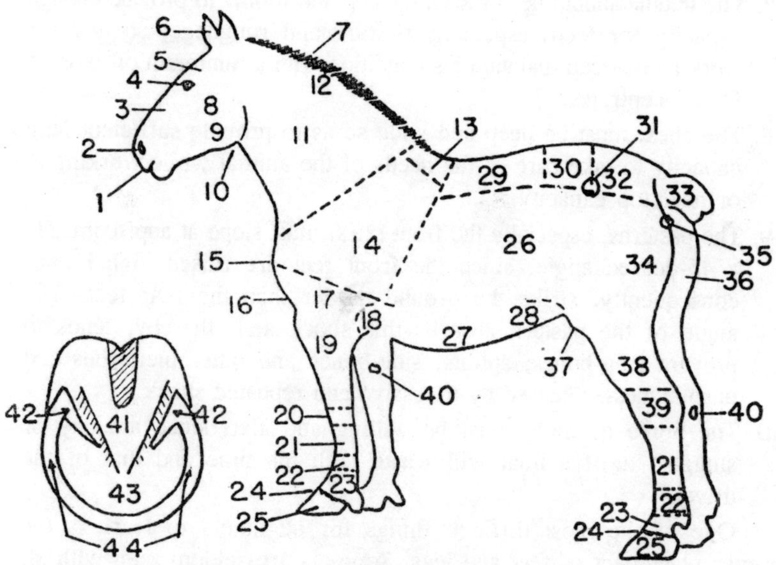

Fig. 9.1. Parts of the horse or mule

1. Muzzle
2. Nostril
3. Face
4. Eye
5. Forehead
6. Ear
7. Poll
8. Cheek
9. Jaw
10. Throat latch
11. Neck
12. Crest
13. Withers
14. Shoulder
15. Shoulder point
16. Breast
17. Arm
18. Elbow
19. Forearm
20. Knee
21. Cannon
22. Fetlock
23. Pastern
24. Hoof head
25. Foot
26. Ribs
27. Belly
28. Rear flank
29. Back
30. Loin
31. Hip
32. Croup
33. Tail
34. Thigh
35. Point of Buttock
36. Quarter
37. Stifle
38. Gaskin
39. Hock
40. Chestnut
41. Frog
42. Bar
43. Sole
44. Wall of hoof

4. The gaskins, quarters, thighs and croup of the animal must be heavily muscled, since these generate the pulling power.
5. The loin and back must be short, broad, and heavily muscled in order to carry the thrust from the rear quarters to the shoulder.
6. The shoulders must be sloping and clean cut so as to give a good seat to the collar and permit an alert carriage of head and neck and a long, easy stride.

7. The middle should be sufficiently deep and roomy to provide enough capacity for feed, especially pasture and roughage, so that the work animal can maintain his condition with a minimum of expense for concentrates.
8. The chest must be deep and wide so as to provide sufficient lung capacity to take care of the needs of the animal while working at or near top capacity.
9. The pasterns, especially the front ones, must slope at approximately a 45-degree angle, since the front feet are raised higher and, consequently, strike the ground harder than the rear feet. The angle of the pastern absorbs this shock and, thereby, tends to prevent ring-bones, splints, side-bones and other blemishes and unsoundnesses caused by excessive and repeated shock.
10. The horse or mule must be willing and alert since a balky or sluggish draft animal will waste both his time and that of the driver.

One of the most difficult things for beginners to learn is the proper placement of feet and legs. Animals are seldom seen with all four legs squarely in place. Show animals having faulty set to legs or hocks are usually placed so as to make these defects less obvious. Once the proper placement of feet and legs is learned, however, it is relatively easy to spot defects. As an aid to learning the correct placement for the feet and legs of a horse or mule study Figure, which shows both the correct placement and the most common deviations from this.

Ideal Draft Animal

Even though horses and mules do not closely resemble each other, they serve the same purpose; their ideals in size, conformation, quality, and action are very similar. The main difference lies in the closeness to which each approaches this ideal. Some mule men contend that a mule that looks too much like a horse lacks the durability expected of a mule. There is some truth in this argument. However, in both the sale ring and the show ring the mule is measured against much the same standard of perfection as the horse. He is usually most deficient in quality of head; muscling over the back, loin, and croup; fullness in the rear quarters; size and quality of feet and bone; and set of hind legs.

Size

The ideal size for a draft animal will vary with the section of the country, the size of the farm, and the job that is to be done. The

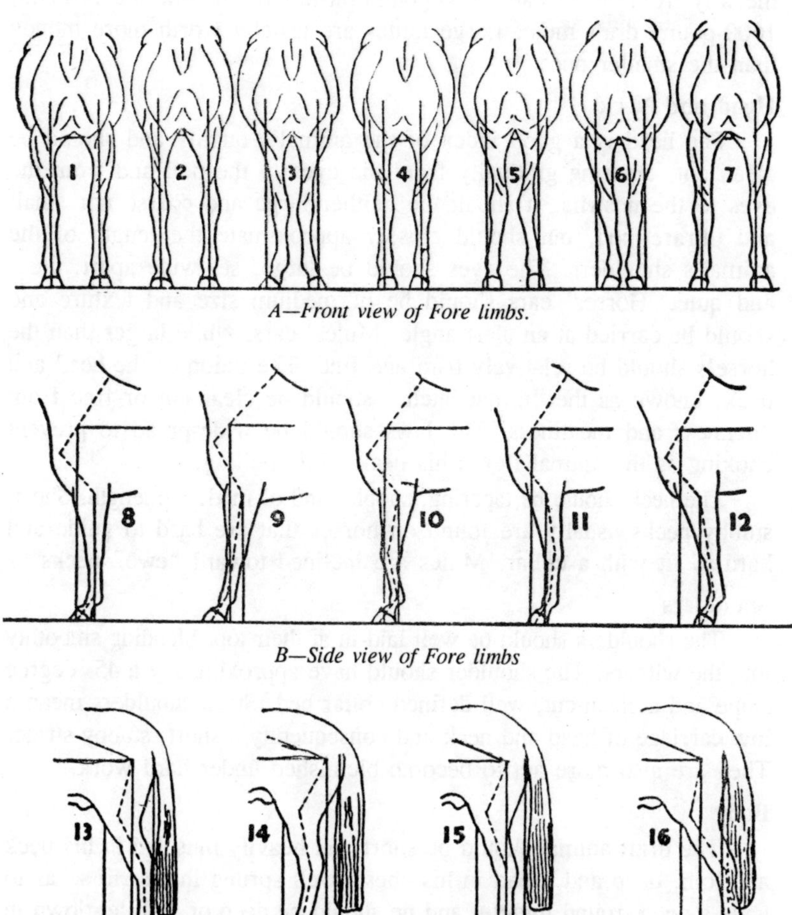

A—Front view of Fore limbs.

B—Side view of Fore limbs

C—Side view of Hind limbs

Fig. 9.2. Placement of feet and legs of a horse or mule.

extra-large, clumsy horse that was popular in the North some 20 years ago, has now been replaced with a 16-hand (a "hand" is four inches), 1600-pound horse that is no taller from the sole of his foot to his chest floor than from his chest floor to the top of his withers. This newer type is more active, handier with its feet, and easier to keep fit.

Southerners prefer even lighter horses and consider a 1000 to 1400 pound horse to be sufficiently large. Mules vary in weight all the way from the 12-hand, 600-pound pit-mining mule to the 17-hand, 1600-pound draft mule. Large mules are usually worth more money than the smaller ones.

Head and Neck

The head is a good index of the animal's quality and should be clean cut, tapering gradually from the eyes to the poll and from the eyes to the nostrils. It should be neither large and coarse nor small and ultrarefined, but should closely approximate the length of the animal's shoulders. The eyes should be large, set wide apart, clear and quiet. Horses' ears should be of medium size and texture and should be carried at an alert angle. Mules' ears, while larger than the horse's should be relatively trim and fine. The union of the head and neck, known as the "throat latch," should be clean-cut or free from thickness and meatiness. The jaws should be widespread to prevent choking as the animal flexes his neck while pulling.

The neck should be tapering, supple, and of moderate length. Short, stubby necks usually are found on horses that are hard to guide and hard to fit with a collar. Mules are inclined toward "ewe" necks.

Shoulders

The shoulders should be well laid-in at their top, blending smoothly into the withers. The shoulder should have approximately a 45- degree slope and a clean-cut, well-defined collar bed. Steep shoulders mean a low carriage of head and neck and consequently a short, stubby stride. They are also more apt to become blemished under hard work.

Body

The draft animal should be short and heavily muscled in his neck and loin, deep and broad in his chest, well sprung in his rib so as to have a deep, round middle, and he should be deep or well let-down in his hind flank. These characteristics insure ample lung and stomach capacity as well as a strong top for heavy pulls. Light-middled horses and mules are hard keepers. Animals with sharp backs and narrow lions lack the necessary muscling to stand up under heavy work.

Rear Quarters

The hindquarters of the draft animal generate most of the pulling power. They should therefore be wide, deep, and heavily muscled. The croup should be long, broad, and gently sloping. The thighs should be wide, deep, and heavily muscled. The gaskins and stifle should show heavy, bulging muscling. The mule is most commonly deficient in the rear quarters.

Feet, Legs, and Action

The proper placement of feet and legs has already been discussed. The feet should be large, round, sloping from front to rear, and wide at the heels. The horn of the hoof should be dense, smooth, and free from cracks. The rear feet are not as side at the heel as are the front feet. While most mules have relatively small, narrow feet, larger feet with quality are preferred.

The pasterns should slope at approximately a 45-degree angle and have sufficient length and elasticity to absorb the shock and to permit an easy gait. Short, upright pasterns go with a jerky gait and unsoundnesses, especially of the front pasterns and legs.

The cannon bones are the best place to observe quality of bone. The bone should be of medium size, dense, and flat. Knees should be deep from front to rear, wide as viewed from the front, and tapering gradually into the leg.

The hock is of outstanding importance since this is the lever used to transmit the power of the rear quarters in propelling the animal forward. The hock should be wide from front to rear, deep from top to bottom, and wide at the front. It should be clean-cut and free from blemishes and unsoundnesses.

The draft animal should have a long, straight, snappy yet easy stride. He should lift his feet high enough to clear obstacles, but not so high as to be wasteful of energy. Throwing the feet outward as they are moved forward is called "paddling". Exaggerated paddling in high-going horses is called "winging". "Interfering" is striking one foot against the other as it is moved forward. Moving forelegs around each other and planting them in a single, straight line is called "winding" or "rope-walking." Striking a front shoe or foot with a rear foot is called "forging." "Rolling" is shifting of weight from side to side to keep balance—a defect found especially in excessively wide-chested horses. "Pounding" is hitting the ground too hard, whereas a quick, high, but short stride is described as being "trappy." Traveling wide at the hocks is a serious fault, since an animal so affected springs out at the hocks under heavy pull and punches himself in the belly with his stifle at the trot.

Common Blemishes and Unsoundnesses

A *blemish* is an imperfection or injury not serious enough to interfere with the serviceability of the animal. An *unsoundness* is an imperfection of form or function that is caused by inheritance, injury, overwork, or misuse and that interfere with the serviceability or usefulness of the animal in a way that greatly lowers his value. Work stock are usually sold with or without guarantees as to soundness as follows:

1. "Sound"—guaranteed free from any unsoundness or serious blemishes.
2. "Serviceably sound"—showing only blemishes or unsoundnesses that do not affect his usefulness.
3. "At the halter"—no guarantee as to soundness, disposition, or any other characteristic.

Unsoundnesses vary as to their effect on the usefulness of a horse or mule. It is usually accepted that the following constitute show-ring disqualifications: lameness, blindness, bone spavin, stifled, stringhalt, ringbone, one testicle or abnormal testicles in a stallion, and bad wind.

The following defects are recognized as ground for more or less serious discrimination: sidebone, bog spavin, thoropin, capped hock, cocked ankle, poll evil, filled hock, curb, fistula, umbilical rupture, and stocked ankles.

Definitions of the more common blemishes and unsoundnesses are given:

Bone spavin—a bony growth usually found on the inside lower part of the hock joint.

Bog spavin—a soft swelling located at the front inside area of the hock.

Thoropin—a soft swelling similar to a bog spavin but located in the depression just in front of the point of the hock. Pressure on one side will cause the swelling to bulge out on the other side.

Curb—a swelling at the rear of and toward the bottom of the hock.

Ringbone—a bony growth around part or all of a pastern.

Sidebone—a prominence and hardness at the hoofhead caused by ossification of the lateral cartilage.

Splint—a bony growth lying along the inside of the front cannon bone below the knee.

Stifled—dislocation of the patella of the stifle joint; usually incurable and crippling.

Stringhalt—a nervous disease resulting in a jerky, abnormally-high action of the hocks.

Fistula—a deep seated abscess at the withers.

Poll evil—similar to fistula but located at the poll.

Heaves—a respiratory disease characterized by wheezy and jerky breathing and a short hollow cough.

Blindness—a result of injury, infection, or "moon blindness" causing loss of sight in one or both eyes.

Capped hock—a prominence at the point of the hock usually caused by bruising.

Cocked ankle—usually found on rear legs with unusually steep pasterns caused by shortening or inflammation of the tendons and resulting in a short, stubby and frequent stumbling.

Market Classes and Grades of Horses and Mules

There are no up-to-date official market classes and grades of horses and mules. They vary from market to market and from year to year. The following represents an average classification for the country as a whole.

Draft Horses

Draft horses surpass the other classes of work animals in conformation, weight, and quality. They are deep-bodied, short-legged, thick, heavily-muscled, and of good quality.

Chunks

These are similar to draft horses, having approximately the same conformation and quality, but less size. They are used principally for farm work.

Wagon Horses

These are similar to chunks in size, but are somewhat more leggy and possess more quality and speed. They are used principally in cities on milk, bakery, and other delivery wagons.

Driving Horses

The demand for driving horses has been practically eliminated by motor vehicles. The remaining demand is principally for pleasure and show purposes. Both outlets expect and will pay for quality. Roadsters and principally of Standardbred breeding and possess quality and speed. Heavy harness horses are principally of American Saddle Horse or Hackney breeding. They are a somewhat more stylish horse than the roadster and possess higher but shorter action. They are, therefore, not so fast as the roadster.

Saddle Horses

These vary in quality from the western bronco to the finest of 5-gaited saddle horses, and in size from the upper pony limit of 14.2

hands to the large, upstanding hunter of 17.2 or better. Light hunters and jumpers are usually of Thoroughbred breeding, whereas heavy hunters and jumpers often carry a cross of Percheron blood.

Ponies

This class arbitrarily includes anything under 14.2 hands in height. They vary from the chunky, rather plain-legged children's pet, to the clean-cut, stylish show pony.

Classes and Grades of Mules

Mules are classified according to their use and graded according to their quality and value. A short description of each class follows:

Draft mules are the largest of all mules and are used for teaming, contracting jobs, and lumbering. These should be of drafty type, good quality, and straight and snappy action.

Farm mules are usually of somewhat poorer type, quality, and finish than are draft mules. They are purchased for use as farm work stock principally in the central and southern states. Mare mules are preferred.

Sugar mules are used on southern sugar plantations. They are between the draft and cotton mules in size, type, and quality. Mare mules are preferred because of their higher quality and more active dispositions.

Cotton mules are purchased for use on the cotton plantations of the South. They are definitely lighter and less drafty in type than the sugar mule. The present trends in southern farming call for larger and better mules than were used a number of years ago.

Mining mules vary greatly in size and weight according to the type of mine work they are to do. They should be of compact, drafty type. Geldings are preferred because of their more sluggish dispositions. Mules are well adapted to use in mines because they are quieter than ponies and less apt to be injured by misuse.

Factors Affecting Marketing Value of Horses and Mules

Class. The market class for which an animal can qualify will determine to a large extent his value. It can be seen that since 1935 saddle horses of even medium quality have outsold the average draft horse, wagon horse, or chunk.

Age. The value of a horse or mule also depends on its age since they may be either too young to work or too old to have much serviceable life left.

Sex. Mares are preferred to geldings wherever the buyer plans to raise foals for replacements or for sale. Mare mules are also preferred, except in mines, because of their higher quality and more alert, active dispositions.

Conformation. The conformation of a horse or mule greatly affects its serviceability and therefore its value. Narrow, upstanding, light-muscled draft horses sell at a considerable discount. So do low-headed, coarse, stubby saddle horses.

Quality. Quality of bone and feet is especially important since it is an indication of durability. General quality is closely related to price.

Colour. Preferences differ, but generally speaking the solid and darker colours are preferred. In teams the colours should be matched.

Temperament. Any evidence of vices or skittishness detracts from the value of an animal. However, a spirited and energetic animal of tractable disposition is preferred.

Training. The average buyer wants a well-trained animal.

Condition. Fat covers a multitude of defects. It also indicates that the animal is healthy and has been well cared for.

Style. This is especially important in saddle horses. A little work in grooming and trimming often pays big dividends.

Action. The action of a horse or mule is closely related to its serviceability. Those with short, low, sluggish, or crooked action can do less work, give a poor ride, and often become unsound due to physical defects that cause undue strain on joints or bones as well as poor action while so doing.

Soundness. An unsound or even a blemished animal should sell at a considerable discount since most unsoundnesses affect an animal's usefulness in a greater or lesser degree.

10

PREVENTIVE MEDICINE

In the developed countries certainly, and to an encouraging extent in the countries of the third world, many of the devastating animal epizootics have either been eliminated, e.g. bovine tuberculosis, brucellosis, rinderpest, pleuropneumonia, swine fever, or they and other serious infections are being contained by the use of vaccines, whose efficacy is being steadily improved.

Nevertheless, there is a large difference between the productivity obtained from the average farmer from all types of livestock and the level consistently obtained by the top one-third of those producers keeping similar animals under the same system of management. The Meat and Livestock Commission in the U.K. has been publishing the results of such comparisons for many years, and show that the extent of this greater profitability ranges from 10 to 20 per cent. This raised farm income has been shown to be due, not only to a more constant avoidance of loss from disease, but from attention to various aspects of animal husbandry, an awareness of cost-saving methods (such as the forward-buying of fertilizers and feedingstuffs), and the keeping of careful records.

In several European countries during the past few years increasing numbers of the veterinary profession, either as private practitioners, Government officials or teachers within veterinary schools concerned with farm livestock, have been exploring ways of extending and combining disease prevention measures with techniques that assist the farmer to make more profit from his animal production.

To be fully effective, systems of preventive medicine involve the combined efforts of all appropriate advisory personnel working through the farmer and his staff, with whom there has to be regular meetings for planning and assessment. The fact that this comprehensive approach to preventive medicine is steadily gaining acceptance makes it necessary to examine how the husbandry of various species of farm animals can contribute effectively to preventive medicine schemes.

DAIRY CATTLE

In the European Economic Community Countries and in various states of North America and Canada, dairying is the most important of the animal industries, milk producers contributing the greatest proportion of the total value from all farm products.

The number of dairy farmers declines every year but with it goes the continuous enlargement of individual herds. As dairy farmers become fewer, herds become markedly larger. In the U.K., for instance, between 1975 and 1979 it is known from Milk Marketing Board records that nearly 1000 dairy farmers gave up keeping cows each year, but the total number of cows in milk did not decline. During the same period, the average size of dairy herd in the national milk records scheme increased from 72.4 to 81.8 cows.

The common Agriculture Policy of the E.E.C. has, an one of its objects, the safeguarding of the standard of living of the farmers within each of the nine countries, so that the importance of dairying in most of the member states ensures that prices of milk, butter and cheese continue to make production profitable. This, in spite of an overall surplus of low-fat milk, is due to the remarkable spread of the Friesian breed. Much of this excess is processed into skim milk powder, which in the end has to be sold at low prices to countries outside the E.E.C.

There is, however, an overall deficit in butter, most marked in the U.K. A study made by a group at Reading University in 1977 concluded that there should be a change in the breed structure, replacing the low-fat Friesian with Channel Island cows, so that their high-fat milk could contribute to the country's butter shortage. So far, there appears to have been scant heed paid to this advice. With the coming into force of the E.C.C's system of payment for milk quality based upon total solids content, it becomes very important to see that feeding and management are of the right order to permit the cow, especially if she is a Friesian, to express her full genetic potential.

Call-Rearing

The importance of the calf-rearing regime in contributing to the life-time productivity of a dairy cow has been studied. It has been made clear that the level of feeding can markedly influence the onset of puberty and that the age of first calving dependent upon it often has a direct relationship to the animals' subsequent production capability.

If healthy calves are fed excessively, a proportion of them can be successfully mated as early as 12 months. Such early-mated heifers are likely to have a higher than normal incidence of calving difficulties followed almost certainly by a low level of milk production. It is thought that this is, at least in part, due to the excessive quantity of inter-lobular fat laid down in the udder as a result of abnormally high energy intake. Subsequent lactations of these cows show greater variation than normally occurs.

If illness or under-feeding delays the beginning of ovarian activity, making first calving later than usual, there is a varying degree of loss of income. If subsequent nutrition is optimal and disease is avoided, life-time production can be satisfactory.

The heifer replacement calf has two periods of disease hazard through which it must pass. During the pre-weaning and immediate post-weaning phase because it is 2-4 months before the calf has a fully functional active immunity-producing mechanism, a series of bacterial and viral infections of respiratory and digestion systems may be encountered. The mortality rate during this young calf stage in many countries is between 6 and 8 per cent.

Intestinal infections

Enteric infections due to viruses, *Escherichia coli*, mycoplasmas, often follow indigestion caused by overfeeding the milk replacer during the first 5–6 weeks. It is believed that the overfilling of the abomasum prevents normal milk-clot formation and the subsequent raising of the pH in the small intestine to higher than normal levels encourages the growth of pathogenic microorganisms. It is also a matter of the greatest importance in the raising of calves that the quality of the skim milk powder that makes up 60 per cent of most milk replacers, is manufactured by the low-heat spray process, designed to avoid denaturing the milk proteins. When cheaper milk powder with damaged protein is used, hard and larger clots are formed in the calf's stomach. Careful mixing of the replacer powder with water and, if twice-daily bucket feeding is used, giving the milk at 41°C, along with strict hygiene, will all help to avoid digestive upsets.

As soon as there is any sign of diarrhoea, the amount (but not the strength) of the milk replacer must be halved for one or two days until the calf recovers.

Respiratory infections

Both the incidence and the severity of respiratory infections are related to the system of husbandry used in raising the young calf. Ample ventilation without any chance exposure to draughts has become well recognized as essential. The question of an optimum temperature appears not to be so critical as it is to provide an environment that causes a minimum degree of stress.

Veterinarians have possibly been too prone to give their support to a system that keeps the very young calf in a separate slatted crate for the first few weeks of its life, the belief being that, in this relative isolation, it would avoid respiratory infections. Calves are of course by nature gregarious animals who delight to move freely about and play with their fellows. Evidence now seems to show that when small groups of calves of similar age are allowed to live and play together in well-strawed pens, disease can be kept at a low level.

After weaning there follows a 'hardening-off' period to about 12 weeks of age, during which calves of similar age and size are run together in well-ventilated pens at air temperature but given cereal supplementation of good quality roughage.

Internal parasitism

It is customary in temperate countries to permit replacement heifers, and calves destined for beef, to graze improved pastures during the spring, summer and into the autumn of their first year. There is need in many temperate countries to vaccinate calves in the early spring of their first year against lungworm. Two subcutaneous injections of a preparation containing irradiated third-stage larvae of *Dictyocaulus viviparous* given in early spring have been used for many years and provide an effective degree of protection. It is at this time that they are likely to pick up from the pasture high levels of infective larvae of various pathogenic intestinal worms. The particular species or combination of species that cause most trouble tend to vary between different countries, but the most dangerous for cattle of all ages are *Ostertagia* spp. and *Trichostrongylus* spp. Improved knowledge of the epidemiology of internal parasitism, associated with increasing use of the pepsinogen test as a diagnostic aid has meant that strategic drenching can be undertaken.

In order to eliminate the developing fourth-stage larvae that have been picked up from the pastures during spring grazing, it has been widely accepted that calves should be drenched during the second week of July; with a second drenching with a benzimidazole anthelmintic or levamisole at the time of housing for the winter. This latter occasion is for the purpose of preventing the development of dormant or protected larvae of *Ostertagia* which, as mature egg-laying worms in the yearling's intestine, will contaminate the pastures in the following spring. At the yearling stage, an age immunity is likely to have developed; although in meteorologically favourable areas, it is desirable to drench these older cattle also.

Recent evidence from surveys in the U.S.A. and in part of the U.K., indicates that increased milk production can be obtained if cows are given anthelmintic treatment in the dry period before calving. This is due to the fact that in favourable situations ostertagiasis may occur sporadically, leading to clinical effects.

Rearing of Yearling Heifers

In temperate climates there are two peaks of calving, in the autumn and the spring, with most born in the autumn. Since whole milk is consumed immediately following pasteurization, a marketing system in industrialized countries has evolved that gives price encouragement to cover the higher costs of feeding cows inside during the winter months. Further advantage inherent in autumn calving for the production of replacement heifers resides in the fact that they will reach the usual mating age of 15 months near the beginning of what is regarded as the Artificial Insemination Centre's mating season, so to calve for the first time in the autumn of the following year.

By the beginning of winter, heifers being kept for herd replacement will therefore either be approximately one year or 8–10 months old. In many dairying countries this will mean that they will be brought off pastures and fed inside during the following four months. If growth is to be uninterrupted, care must be taken to see that supplementation of the declining nutritional value of the pasture is undertaken for some weeks before they are brought inside in October/November. This will be achieved by giving 0.5–1 kg of cereal (crushed barley, oats, wheat, maize) along with a mineral mix or lick. The aim during the inside winter feeding period is to keep costs to a minimum yet achieve a modest continuation of growth with daily weight gains of between 0.5 and 0.75 kg/day. The cheapest way to do this is to permit ad lib feeding of good quality grass silage (DM 25–30, CP 16, energy at 11 MJ per kg DM). In areas where good quality grass or legume hay can be made, this fed in generous quantity will permit similar weight gains.

Mating of Heifers

Mating or artificial insemination will normally be undertaken when the heifer reaches 15 months old and is having normal oestrous cycles of 21 days. It is of the utmost importance that observation to detect signs of oestrus is properly organized because, in some heifers, particularly amongst zebu breeds, signs of heat may be very meagre and thus hard to detect—so called 'silent heat'. There is now available and increasingly used, a method of giving two injections of a synthetic prostaglandin 11 days apart, followed by artificial insemination 76 hours after the second dose. Some veterinarians favour two inseminations, the second given 24 hours after the first. The evidence indicates that careful use of this technique not only improves the percentage of conceptions but, by obtaining calving approximately 30 days earlier, the extra income after deducting the cost of feed, injections and extra insemination at present-day prices can be about £27 per heifer.

Where the use of prostaglandin and batch mating by artificial insemination is not undertaken, it is a common practice to run a bull with the heifer group, although artificial insemination is likely to be routinely used with the older cows. In order to avoid the possibility of difficult calving, many dairy farmers use a beef bull with heifers, such as a Hereford or Aberdeen Angus. The calf is not only smaller at birth than the pure Friesian/Holstein, but is more readily sold for eventual fattening for beef.

Cows

The areas of preventable loss in lactating cows can be grouped under the following headings:

I. Diseases from infections of (1) the udder, and (2) the feet.

II. Production diseases from nutritional imbalances.

III. Reproduction disorders.

Diseases from Infections

Infections of the udder (mastitis)

This remains the commonest and most costly disease of dairy cows the world over. There are many microorganisms that can cause inflammation of the udder tissue, some leading to systemic involvement and general illness, others restricted to the mammary gland. All four quarters may be infected as can be seen by the pain and swelling that results. Occasionally there is blood as well as clots in the milk. Frequently there are the signs of a more chronic type of infection in which only one or two quarters are involved.

Many cases of mastitis will respond favourably to the injection of a broad-spectrum antibiotic but it is often advisable to discover whether streptococci, staphylococci, coliform or mycoplasmas are responsible, and thus be in a position to decide what form of treatment is most likely to be effective. If the assistance of a diagnostic laboratory is available, a milk sample should be obtained. The teat of the affected quarter is carefully cleaned and after swabbing the orifice of the teat with spirit. 10 ml of the fore-milk should be drawn into a sterile bottle. After labelling, this must be sent speedily to a laboratory.

The invading bacteria almost always gain entry via the treat canal. It is obvious, therefore, that care must be taken to avoid, as far as possible, spreading infection when the milking machine cluster is moved from cow to cow. Early detection of mastitis is accomplished by using a strip cup to examine the fore-milk or an in-line indicator may be used. Teats should also be dipped into or sprayed with an iodine or hypochlorite disinfectant immediately after milking.

There have now been many years of experience in the use of a simple, practical routine, first worked out at the National Institute for Research in Dairying and adopted by the Ministry of Agriculture and Milk Marketing Board. It is a fact that losses from mastitis can be controlled, if not eliminated, by constant adherence to this five-point programme:

(a) *Teat dipping* (or spraying) immediately after the removal of teat cups. A mixture of glycerine and iodophor is most commonly used.
(b) *Culling of chronic cases*. This reduces the risk of spread of infection.
(c) *Clinical* cases must be promptly and correctly treated.
(d) *Dry cow treatment*: all cows must have a broad-spectrum antibiotic inserted into their udders at the end of lactation.
(e) *Machine testing*, regular examination and testing of all parts of the milking machine to see that teat liners, pulsation ratio and rate, and the vacuum pressure are correct. This helps to prevent teat damage.

The detection of early infection of the udder or the sub-clinical chronic case is assisted by using the Californian mastitis test. It depends upon detecting the increase in white blood cells in the milk that always accompanies udder infection. A small quantity (2 ml) of the fore-milk is mixed with 2 ml of a 3 per cent solution of alkylaryl sulphonate. The anionic detergent reduces surface tension and the cells clump together. The degree to which this occurs indicates the seriousness of the infection.

The importance of general hygiene in reducing faecal contamination of the teats and the likelihood of physical injury has to be stressed. The construction of cubicles or kennels and the type of bedding provided, the design and maintenance of any slatted area, as well as the daily removal of slurry from feeding and collecting points, care in udder washing, all must be continuously monitored if mastitis is to be contained. That these measures are effective in reducing sub-clinical mastitis is seen by the fact that surveys in the U.K. in the 1960s showed that 50 per cent of the cows in the national herd were infected. From a survey carried out in 1976/77 it appears that it was then down to 32 per cent. It was, however, concluded that the number of clinical cases of the disease had not decreased.

Diseases of the feet

In the temperate dairying countries periods of high rainfall occur. As a result, the cows' feet get caked with mud and sometimes stones become stuck between their claws, the resulting abrasions permit the entrance of pathogenic strains of bacteria (e.g. *Fusiformis necrophorus*) that are likely to cause abscess formation. Effective treatment may be given along several lines, but prevention is not an easy matter. The avoidance as far as possible of having mud dry in the interdigital cleft can be achieved by regularly putting cows through a foot-bath to which formalin has been added to make a 10 per cent solution. Loose stones should not be in the cows' path.

Where there is an unusually high incidence of inflammatory foot lameness, the possibility of a marginal zinc deficiency should be considered. This may arise from too high an intake of calcium making some of the dietary zinc unavailable.

It is believed that in the U.K. about 8 per cent of the dairy cows become lame every year and that laminitis and its complications are responsible for about half of this total. This inflammation of the sensitive laminae of the hoof is considered to be a metabolic disorder primarily due to excessive production of lactic acid in the reticulo-rumen, due to the sudden ingestion of a large proportion of grain in relation to the forage intake.

It occurs mainly during the winter period when stock are housed and, more frequently, in heifers. This appears to be due to a combination of factors, amongst the most important being restriction of movement and unaccustomed standing on a cement surface, imbalance in feeding due to full intake of concentrates when in the milking parlour, but competition from older and heavier cows afterwards at the silage face or feed trough greatly reducing the intake of forage. A frequent result of an acute attack of laminitis is a chronic condition of the hoof that leaves the sole flattened and liable to bruising and penetration, sometimes to white line separation and abscess formation, or to sole ulceration, always to horn overgrowth.

Preventive measures therefore, must aim at gradual change of feeding and housing, of making special provision to see that all dairy stock get a balanced diet. Here the growing tendency to do away with feeding concentrates in the milking parlour in favour of complete diet feeding is helpful. In any herd that is troubled by this disease, the addition of 1 per cent sodium bicarbonate to the mixed ration may also help. Where silage is self-fed or dried forage fed in troughs there must be adequate space allowance. All overgrown hooves must be trimmed.

Production Diseases

This is the name used to describe illness in dairy cows associated with malfunctioning of digestive or lactational processes. They were for a long time known as metabolic disorders. They include parturient paresis of milk fever, ketosis or acidosis and acute hypomagnesaemia or grass tetany.

There are severe physiological demands made upon the modern high-producing cow. She has to convert the greater part of the energy contained in the food eaten into milk during a lactation of 305 days. The peak of her lactation curve occurs at 6–8 weeks after calving and during the early part of these 45–56 days, she has a lowered appetite and an increasing milk demand. She will be unable to absorb enough energy to produce to her genetic capacity and at the same time maintain her weight. This weight loss must be kept to a minimum, and for the average Friesian a total reduction of 25 kg is regarded as tolerable. The important thing is to see that the intake of metabolizable energy is sufficient during the later months of declining lactation to make up the lost weight. The aim must be to have each cow at her optimal weight before the dry period and the time of calving comes round again, hopefully at about one year from her last parturition.

It is not surprising that the time of greatest hazard for the cow is from calving until the lactation peak. It also happens for many cows that at some point during this period, they will experience a change in feeding routine. If they are autumn calvers it will be when they come indoors on to winter feed. If spring calvers, it will be as they are turned out to graze the new pastures.

Milk fever

This occurs at the onset of lactation, that is, within the first three days of calving, and is commonest in older, heavy-milking animals, when it will occur within 24 hours of calving. The affected cows cannot stand and as she lies, frequently has her head turned towards the flank. In severe cases, she will become comatose and die unless treatment is given promptly. The cause is an inability to mobilize enough calcium from the large quantities stored in the bones to put into the milk that is suddenly required. Samples of blood taken at this time are likely to show low serum calcium values.

Our present understanding of how this state of affairs comes about is that the intake of calcium towards the end of lactation and in the dry period following it is more than the cow needs, thus giving rise to a predominantly calcitonin-mediated effect on the intake/output of calcium. The function of calcitonin, a hormone produced by C-type cells within the thyroid gland, monitors the output of calcium from the kidneys and opposes the bone-calcium mobilizing function of the parathyroid gland. The greatly increased demand for the mineral at the onset of lactation can only be met if parathormone from the parathyroid gland takes over quickly enough to allow the bone calcium stores to be used.

In a proportion of cows, this cannot be achieved fast enough, particularly if the dry period ration has contained a lot of calcium, or if cholecalciferol (vitamin D) stores have not been ample. This vitamin plays a decisive role in the transfer of calcium from the digestive tract into the blood.

Clinically affected cows generally respond promptly to 8 g of injected calcium in the form of calcium borogluconate. Prevention consists in restricting the intake of calcium during the last few weeks of pregnancy from the usual 25 g per day to 8–10 g per day, but increasing it again from calving onwards. If the calving day is accurately known, an injection of 0.5 μg per kg body weight of the synthetic 1-α-hydroxycholecalciferol during the 24 hours before calving, will significantly increase the uptake of calcium from the gut. Although a success rate of 88 per cent has been reported with 60 cows on one farm by this method, a more extensive trial that was carried out by Sansom and Allen (1981), involving 1200 cows in 25 herds, showed that the efficacy of prevention varied markedly from nil on one farm to 74 per cent on another, with an overall effectiveness of 32 per cent.

Ketosis

The name is derived from the fact that when the blood glucose falls to a low level (hypoglycaemia), certain normal products of carbohydrate metabolism, β-hydroxybutyric acid, acetoacetate and acetone, collectively known as 'ketone bodies' are unable to proceed to help in the formation of glucose because of the failure of oxaloacetic acid to be formed. It most often occurs three to six weeks after calving, and although not many cows die, severe loss of production and weight inevitably follow.

The disorder occurs because of an imbalance between the intake of metabolizable energy and the output involved in heavy milk production. It may occur in an acute or in a more chronic form. Indeed, evidence suggests that a sub-clinical form, shown by periods of lowered appetite and reduced milk yield, may be as high as 80 per cent in intensively managed cows kept indoors. This form of the disease is often caused by some other illness, for example, acute mastitis, onset of foot infection or endometritis, all causing an interruption in normal food intake. Not infrequently the cause is the inadequacies of the ration being fed or, more precisely, the quantity and quality of the energy content of the ration is too low. An analysis of the forage and a simple calculation of the daily amount of metabolizable energy being obtained related to the cow's weight and milk yield will reveal how to put the matter right.

The typical clinical signs of the acute, spontaneous form of ketosis are the sudden refusal to eat concentrates with the following sharp decline in milk production. Body condition is rapidly lost, the animal is dull, sometimes shows nervous symptoms and passes dry glazed faeces. These cases will respond to the giving of glucocorticoids through the repression of ketogenesis and stimulation of gluconeogenesis. Anabolic steroids may be given to stimulate appetite. Certainly, intravenous glucose solution (500 ml of 40 per cent solution) should also be given. Further support may be provided by giving, orally, propylene glycol or sodium propionate (250–500 g per day, in two doses for 5–10 days).

The preventive measures advocated by the Institute for Research in Animal Diseases at Compton, England, are effective and include:

(a) Avoidance of cows become overfat during the last part of lactation and the dry period.

(b) Restrict any 'steaming up' to the last month before calving so as to make this part of a continuously rising plane of nutrition up to the peak of lactation.

(c) The ingredients of the ration should be kept constant. Any necessary change must be made very gradually.

(d) There must be a correct balancing of nutrients, with the protein not exceeding 17 per cent and, if possible, the cereal component of the concentrates should include ground maize, some of whose energy as well as protein will escape rumen fermentation and yield glucose in the intestine. After peak lactation, a cheaper concentrate mixture can be given.

(e) All cows must have adequate time and trough space to eat their concentrate allowance especially before lactation peak.

(f) All cows must have an opportunity for exercise.

(g) The greatest importance should be put upon the quality and thus the palatability of the forage fed. A grass silage with a butyric acid fermentation is likely to be poorly eaten as well as being ketogenic.

Hypomagnesaemia

The incidence of grass tetany, the acute form of hypomagnesaemia, is not so high as milk fever but the mortality rate is greater. The disorder commonly appears after the milkers are turned out on to spring pastures Some of the latter may have a marginally low content of the mineral, but it has long been realized that there are other precipitating factors to account for the fact that only a few cows in most herds will show clinical signs of the disease. They include individual variation in the rate of passage of digesta through the rumen (the main organ of absorption), usually much increased when cows begin grazing young pasture, as well as much variation in digestive and absorptive capacity of food ingredients that may interfere with magnesium uptake.

The disorder causes hyperaesthesia or a general nervousness and overreaction to any stimulus. The cow may become recumbent and show twitching of muscles. The appropriate treatment is to administer a solution of magnesium sulphate, often combined with calcium borogluconate because many cows show a degree of hypocalcaemia as well.

The effective preventive measure is to feed additional magnesium to the herd. Most commonly, this is by incorporating calcined magnesite (magnesium oxide) in the concentrate ration to the extent of 50 g per head per day prior to and during the early spring grazing. Equally effective, is to top-dress the pastures just before grazing with magnesite at a rate of 31 kg per hectare.

Biochemical Aids in Preventive Medicine

Analysis of feedingstuffs

Most of the ingredients used in making up the concentrate part of any ration fed to ruminants have a limited range of values and are themselves generally few in number. Cereals make up most of these mixtures to the extent of 70-85 per cent, and the median figures given in up-to-date tables for maize, sorghum, wheat, barley and oats can safely be used. The protein content of different samples of wheat and barley can vary widely, e.g. from 8-15 per cent, so that it may for particular purposes be necessary to check on the precise composition of the cereals being used. Tables may safely be used to calculate the contribution being made by any oil-seed meal that may be included as a protein concentrate.

The important ingredient that is likely to have a wide degree of variation is the forage or roughage. Whether hay or silage is being fed, it is exceedingly difficult and unreliable to depend upon observable characteristics. A chemical analysis of a truly representative sample must be made and then it is a simple matter to apply prediction equations to give the estimated metabolizable energy value :

$$\text{ME (MJ/kg)} = 14.3 + 0.017\text{CP} - 0.019\ \text{MADF}$$

(CP = crude protein;
MADF = modified acid detergent fibre)

Also, by using in vitro methods a fairly reliable way of determining the digestibility of the organic matter (the 'D' value) in a forage has been routinely used.

Analysis of blood

For a long time it has been customary to confirm a tentative diagnosis of mineral deficiencies by taking a series of blood samples from the herd and have the appropriate estimations made. With the availability a decade ago, of automated analytical techniques for many important biochemical components of the blood, the notion of regularly monitoring their levels in a proportion of the milking herd was advanced. The scheme was known as the Compton Metabolic Profile test after the Institute for Animal Diseases where it began and where the important range of normal values were established. It involved collection of blood samples from seven high, seven medium producers and seven dry cows in the herd on two occasions, autumn and spring. Several commercial laboratories provided this service, sending to the farmer and his veterinarian a print-out of values for a large number of blood components, at the same time showing the accepted normal limits for each. At the present time, most veterinarians undertaking preventive medicine services prefer to offer their clients a closer surveillance of herd health by taking blood samples from a proportion of cows every month and doing the estimations of a few key values in their own practice laboratories.

As estimation of blood glucose gives a good guide to the adequacy of the energy intake; estimations for blood urea and albumin will show if there is sufficient protein in the diet; examination of the blood for levels of the major minerals can usefully check that the dietary balance and calculated intakes are within correct limits.

The commonest trace element deficiency is of copper. Caeruloplasmin estimation from a blood sample gives a fairly reliable indication of the available copper, but regular inspection of the herd should reveal the first clinical signs of a deficiency as quickly, in change in coat colour and effect on milk yield.

Success in preventive medicine schemes depends upon being able to discern the inter-relationships between the soil, pastures and fertilizer applications, feeding practice and the system of management operating on a particular property.

Reproduction Disorders

In addition to the large demands that the daily increasing production of milk makes upon the freshly calved cow, she will be expected to be having fertile oestrous cycles within two months. For official milk records, the lactation length is 305 days, to be followed, if possible, by a 60-day dry period. It is therefore clear that it is necessary to have each cow calving again by the year's end if milk production from the herd is to be maximized. The Milk Marketing Board in the U.K. showed some years ago that the loss of milk, and hence of income for the dairy-farmer, amounted to an average of £4.4 per cow per year (costed for a 60-cow herd), when the service interval increased from 85 days to 106 days, i.e. one oestrous cycle. If the calving interval for a 60-cow herd is 13.5 months, there was at that time (1968) a yearly loss of profit of £10.7 per cow compared to a 12-month calving interval.

Non-infectious infertility

For a number of reasons, it is difficult to have a calving index, taken to be the average length from one calving to the next for the whole herd, that is not in excess of 365 days. The important factors responsible for what is sometimes called non-infectious infertility are the following:

1. The *difficulty of detecting oestrus* in the large herds that are steadily becoming common. This is partly because of the lack of time available to the stockman, especially during winter when cows are inside and have to be wholly fed and much more cleaning done in the standings. Periods of patient observation morning and evening are likely to be much reduced.

   Young cows, after their first calving, especially if they are mixed with older cows, have a low position in the social order and may experience 'silent heat', showing transient or no outward signs of oestrus. Because of difficulties in getting enough time at the feeding trough, their weight loss to peak production may be excessive.
2. *Ovarian dysfunction* takes two forms. There may be a persistent corpus luteum producing progesterone to stop normal cycling. This may well be the sequel to an endometritis immediately following calving, or there may be follicular cysts, sometimes associated with high oestrogen secretion that shows as a varying degree of nymphomania.
3. *Inadequate nutrition* may be the cause of delayed conception because it has been shown that cows losing much weight during the period to peak yield are shy breeders. This, then, is another important reason, in addition to preventing ketosis, that the feeding of a high quality concentrate mixture along with the best forage available, should be started at a low level towards the end of the previous lactation, carried through the dry period, increasing steadily as calving approaches (the old 'steaming-up' period) and on into the vital first part of lactation.

The newly calved cow's appetite is reduced by perhaps 25 per cent, so the required energy level can only be attained by offering enough of a palatable high density concentrate.

It would clearly be helpful if one knew the actual weights of the cows being monitored but as yet few farms have a weighbridge. It is possible, by adopting a system of 'condition scoring' from 1-5, to obtain an objective assessment of a cow to indicate whether it is in a fit state for mating. The area of inspection concentrates upon the sacral (loin) region. In thin cows the transverse processes are clearly seen and the spinous processes of the sacrum easily felt. In cows that are at their optimum weight, or at grade 3, the whole area is rounded and the spinous processes only felt upon exerting pressure.

The gestation length of the dairy cow is approximately 283 days, so to be sure of a calving interval of 365 days means that conception must have occurred by 82 days from her last calving. The likelihood of a fertile mating increases to a maximum between 60-80 days post-calving; but it is also a fact that conception rate to first service in modern dairy cows is not much above 60 per cent. Thus, in order to maximize the chances of getting this proportion of the remaining 40 per cent of cows in calf when they cycle again in 21 days time, it is desirable to have all cows, including those previously in anoestrus, having a fertile heat period as close to 60 days post-calving as possible.

Importance of Regular Visits by Veterinarian

An effective preventive medicine scheme would therefore mean making regular visits to the farm either weekly for the really large herds or fortnightly, so that all cows can be examined between 46-56 days after calving. At a first visit, the cows reported as not cycling would have a rectal examination made. If the reproductive tract felt and appeared normal, a 'kamar' heat detection pad could be attached to the rump of the cow. If the animal comes into heat, she will permit others to mount her and thus will be marked.

Improving Conception Rates

Persistent corpus luteum

The treatment for anoestrous cows may begin by having a progesterone estimation made on a milk sample. A high figure indicates a persistent corpus luteum. This can be confirmed by a rectal examination by the veterinarian. There must now be the administration of prostaglandin, for this has a powerful luteolytic effect. It may be given as an intramuscular injection or in the form of an implant placed under the skin behind the ear. Some are given with a low dose of an oestrogen, believed to have a stimulating effect upon ovarian function. Mating or insemination can take place at the first oestrus after any implant has been removed or the injected prostaglandin's effect subsides, likely to occur in 3-4 days. There are occasions when cows that have had an endometritis and not responded to the usual earlier treatment with stilboestrol (a synthetic oestrogen) also develop a persistent corpus luteum. These may be successfully treated with prostaglandin.

Follicular cyst

Those cows that have a follicular cyst and show continuous heat are treated with luteinizing hormone plus progesterone (nymfalon) or with a slow-release progestin (synthetic progesterone).

Controlled mating

Two methods are currently being used in herds where conception rates are unsatisfactory. Both depend upon interrupting the normal cycle by administering a progestin along with an oestrogen for 9-12 days.

The latter is needed to help in preventing the formation of a corpus luteum, or to help in the rapid regression if one has been forming. The administered progestin acts as if there were a functioning corpus luteum. On removing the source of the progestin, the blood progesterone levels decline rapidly to be succeeded by ovulation within a few days.

One form, norgestomet with oestradiol valerate, is given as a small implanted pellet, inserted just behind an ear along with an intramuscular injection of both hormones. After 9 days, the implant is removed by making a small incision and squeezing it out.

Another way is to use a progesterone-releasing intra-vaginal device. This is a Silastic coil, an inert silicone elastomer, that is uniformly impregnated with 2.25 g of progesterone. It also has 10 mg of oestradiol benzoate in a gelatin capsule attached to it. The coil and capsule are inserted into the vagina by means of a lubricated speculum. The oestrogen is rapidly absorbed but the progesterone is steadily absorbed over the following 12 days, Cows must be inseminated or mated at 48 and 72 hours after the device has been removed. If, for any reason, only one insemination can be made, this should be at 56 hours after removal.

These methods should be seen for what they are, namely, means of overcoming management difficulties in detecting oestrus in some animals and in obtaining fertile oestrous cycles in cows that have temporary abnormal imbalances of their reproductive hormones. Surveys have shown that in the spring there can be as many as 26 per cent of cows in a herd that are anoestrus.

Such treatments do not increase the fertility level of the herd. Conception rates will be similar or a little less than normal cycling and inseminated cows. The conception rate is likely to be lower because the true anoestrous cow is obviously of lower reproductive efficiency. By the use of these methods, the calving to conception interval should be significantly reduced.

It is known that stress will cause raised progesterone levels. There must therefore be no change in the diet, in the feeding routine or any extra handling of the cows during the time they are being treated. The handling that is necessarily associated with treatment must be quietly undertaken. If many cows have to be inseminated on the same day, a fresh inseminator should be used for each 30 cows, since fatigue significantly lowers the efficiency with which the task is carried out.

Pregnancy Diagnosis

It is important for the dairy farmer to know as soon as possible if his cow is pregnant. Examination *per rectum* by an experienced veterinarian will reveal whether conception has occurred. With heifers this can effectively be undertaken from about the 35th day of gestation but with older cows, an examination at 6-7 weeks is more reliable.

Where facilities are available, and such a service is provided by the Milk Marketing Board of England and Wales, the raised progesterone level in a milk sample taken at the 24th day of gestation indicates, with a reasonably high level of accuracy, whether a cow is pregnant. Negative readings are more reliable. It has recently been shown that measurement of oestrone sulphate in the whey fraction of a milk sample at about 100 days of gestation has a high degree of accuracy. It could thus be used to confirm pregnancy, or, if negative following an earlier positive identification, to reveal embryonic death and resorption.

Records

In all preventive medicine schemes, success greatly depends upon the accuracy and relevance of the information about the animals in the herd that is available Careful record keeping is vital.

There are many systems available, some use wall charts, or there are circular wall boards on which there is space to record service and calving dates for small herds. These useful visual displays must not replace the careful recording of information in a book or on sheets kept in a convenient place in the milking parlour for the herdsman to make a daily record. Each cow should also have its own card on which the following basic data must be entered:

1. Calving date
2. Date of observed oestrus
3. Service of insemination date and note of bull used
4. Date of veterinary inspection for pregnancy diagnosis
5. Date milk samples taken for progesterone estimation and result
6. Date of drying-off and administration of dry-cow therapy used
7. Date of all episodes of illness or injury; treatment given and result
8. Culling and reason

It is possible for the owner and the veterinary surgeon to calculate some useful indices from the data recorded by the stockman:

	Target
1. Calving to first service	65 days
2. Calving to conception	85 days
3. Conception rate	84 per cent
4. Oestrus detection	80 per cent
5. Culling percentage	15 per cent

There is increasing use being made in larger units of on-farm computers for the purpose of recording of information on all animals, with the associated advantage of rapid play-back and visualization.

Beef Cattle

Many of the disorders discussed in the preceding section may occur with beef cattle. Under the single suckler system of production, it is often desired to have as many of the cows as possible calving together during either the autumn or spring.

Controlled Mating

Varying degrees of success in controlled mating are claimed for the use of prostaglandin. Several synthetic forms of the hormone are now being used e.g. cloprostenol ('Estrumate', I.C.I.), dinoprost ('Lutylase', Upjohn). Two injections of these prostaglandins are given at an interval of 11 days. Cows come into oestrus on the third and fourth days and are inseminated or mated. The use of artificial insemination allows the semen of progeny-tested bulls to be used, thus obtaining superior offspring. The conception rate will probably not be so high as with natural mating. A bull is normally turned into the herd in order to pick up re-cycling cows.

Unwanted pregnancies in heifers too young to carry a calf may be terminated if a prostaglandin is used up to 120 days from conception.

What is beyond doubt is the influence of body condition upon the fertility of suckler beef herds. Kept upon the often varying nutritional adequacies of hill grazing, in some seasons it will demand a high degree of managerial skill to have the breeder cows at their optimum weight and all cycling normally when it is decided to turn in the bulls or begin inseminations. To decide upon the right time to begin supplementing the forage with cereals is by no means easy. In some areas, rainfall and temperatures in the autumn may vary widely from year to year. Even when the pastures appear to be highly nutritious at this time, digestibility of the dry matter has almost certainly declined more than one would suppose if it is not possible to weigh the breeder cows, a useful quantitative assessment may be made by using 'condition scoring' as has been described for dairy cows.

Sheep

In addition to the importance of making special efforts to see that energy and protein intakes are sufficient at the critical times in the breeding ewe's yearly cycle, viz. at mating, during the last weeks of pregnancy and during lactation, there are disease hazards against which preventive measures should be taken.

Infections

At lambing and shortly afterwards, the ewe herself and the lambs are prone to succumb to infection with several types of anaerobic bacteria. It is possible to provide satisfactory protection by injection of the appropriate vaccine during pregnancy. Where deaths have previously occurred from more than one of these anaerobic infections it is customary to use a polyvalent vaccine, thus achieving a blanket protection.

Worms

Lambs are susceptible to internal parasitism if they graze pastures that are heavily contaminated with various types of pathogenic nematodes. On properties where there have been sheep for many years, it can be safely assumed that infective larvae will be found on all permanent pastures. It is advisable therefore to give an anthelmintic to all lambs at weaning. They should then be put on hay aftermath, other comparatively parasite-free pasture or a fodder crop to minimize re-infection.

Foot-rot

One of the commonest infections of sheep the world over is foot-rot, due to a combination of two types of bacteria. One is a fairly common organism to be found in the soil of damp places contaminated with faeces (*Fusiformis necrophorus*) and a small spirochaete (*S. penortha*) confined to the sheep's foot lesion. The lameness that follows infection can seriously interfere with an animal's ability to graze. With fattening lambs, this can delay the time taken to attain market weight by as much as 25 per cent.

Treatment depends upon the removal of over-grown and underrun horn of the hoof so that healthy tissue is exposed. The operation should be carried out with the aid of special foot-rot shears and a sharp knife. Each affected foot must then be placed in a 10 per cent formalin solution. The affected foot may be sprayed with other disinfectants contained in an aerosol if only a small number in the flock are affected. Chloramphenicol (chloromycetin) should not be used because of the danger of producing drug resistance against an antibiotic that has special therapeutic value. If many sheep have to be treated, a foot-bath containing formalin (10 per cent) solution should be used.

All the sheep in the flock may then be given an injection of a toxoid vaccine that has been available for some years. High hopes were held at one time for the latter, both as a way to protect sheep from infection and as a means of treatment that could avoid the tedious surgical procedure. So far, the response shows too much variability and for technical reasons it has proved difficult to maintain the antigenic potency of the vaccine. Thus, while the vaccine may be a useful ancillary aid, the elimination of the disease from a flock still depends upon the careful inspection and treatment of each animal where infection exists. Those ewes with chronic lesions should be sold.

Prevention, likewise, depends upon making a thorough examination of the feet of any sheep brought on to the property. If time or availability of labour does not permit this, all introduced sheep should be kept apart from the home flock for some weeks at least to try to be sure that fresh infection is not introduced.

Pigs

Pigs have an admirable facility of being able to adapt to a wide variety of environments and dietary regimes. But it is nevertheless true that with modern strains of hybrid pigs, growth rate is rapid and food conversion efficiency high only when there are no deficiencies or imbalances in the diet and low temperature or over-crowding are avoided.

Starting a Herd

Measures designed to avoid loss must be comprehensive. If possible, the breeding stock should come from an establishment known to be free of enzootic pneumonia and atrophic rhinitis.

In the U.K. and in other European countries where pig production is an important animal industry, it has been considered worthwhile to begin new units with stock obtained by Caesarean section and initially reared in sterile incubators, then in new pig houses or where pigs have been absent for a period and which have been subsequently thoroughly disinfected. In the U.S.A. this system began in the 1950s and was known as the specific pathogen-free method of establishing healthy herds.

With the much larger herds that are to-day more commonly required, such as involved and costly process is not a practical method. A recent development indicates that with the availability of drugs that are effective in eliminating specific infections, a satisfactory method of beginning a new nucleus herd, free from the common pig diseases, might be to combine drug therapy with very early weaning.

Sows were washed, then on the day of gestation placed in a clean portable farrowing house placed in isolation. Then followed a programme of medication using tiamulin hydrogen fumarate to combat *Mycoplasma hypopneumoniae*, the cause of enzootic pneumonia, and trimethoprim sulphonamide combinations (Tribetrin and Tribrissen, Burroughs Welcome) active against *Bordetella*, believed to be the cause of atrophic rhinitis. Both sows and piglets were given these drugs in appropriate form and dosage; the sows from the time of entering the farrowing house until the piglets were weaned at 5 days old. The piglets were medicated each day after birth until 10 days old, i.e. until 5 days after weaning.

It was the piglets that were in good health and had reached 2 kg weight by 5 days that were taken from their dam and moved in an insulated container to an isolated, portable weaner, flat-deck house. Here they were reared to between 5 and 8 weeks of age in groups of about 12 to a pen. Heated floor mats were provided for 2 weeks, with a house temperature of 27°C to start with, gradually being lowered to 24°C. Growout units were simple open-sided barns with plenty of straw bedding, isolated from other pigs. A series of autopsies carried out at various ages showed that they were entirely free from enzootic pneumonia and atrophic rhinitis.

Keeping Free of Infection

The importance of careful insulation of roofs and floors has been discussed in chapter 6. The avoidance of epizootic disease, such as swine vesicular disease and swine dysentery, depends upon the utmost care in the purchase of feedingstuffs as well as being sure about the health of stock replacements. Even the coming on to the property of the lorry collecting pigs for market and its driver can be the means of spreading infection. A special loading area should be provided that has no access to pig houses containing breeding stock. Strict observance of the use of disinfectant foot-baths, use of clean cotton hats and gumboots for all entering a pig house, no visitors, and attendants not to visit other pig units should be routine precautions.

Preventing Perinatal Loss: Night Farrowing

It has been well known for a long time that the source of serious loss is the death of piglets within the first few days after birth. A recent survey reports over 50 per cent of piglet deaths up to weaning occur during the 48 hours after farrowing.

When sows are conditioned to the restriction of a farrowing crate and given careful supervision, the number of deaths at this time can be greatly reduced.

Several workers have noted that there are more sows farrowing at night than during the day, a not surprising observation in an animal that in the wild state is largely nocturnal. It has, more recently, been shown that by the use of a synthetic analogue of prostaglandin $F_2\alpha$, cloprostenol, it has been possible to obtain what is, in effect, a degree of batch farrowing at night.

This resulted in significantly more piglets being reared than when the time of farrowing was not controlled. There must of course be carefully kept records of service dates, since the timing of the administration of the luteolytic agent depends upon knowing the day the sow or gilt is due to farrow.

By giving 175 μg of cloprostenol in 2 ml of citrate buffer by deep intramuscular injection between 18.00 and 20.00 hours, farrowing began during the following night. The mean time for its onset was 257 hours later; with over 90 per cent starting to farrow between 16 and 34 hours following treatment.

The advantage of night farrowing in a large piggery is that the stockman is not diverted by other duties, thus being able to give his entire attention to the sow and her piglets. He can see that there is no suffocation in the foetal membranes and that the newborn piglets get themselves to or are put under the heating lamp immediately after birth to avoid hypothermia and crushing. It is easier to undertake interfostering and, by marking all that suckled, to identify any that require special care and extra colostrum.

This system has permitted over 5 per cent more piglets to survive than during any of the previous three years in one big piggery.

Worms in Pigs

With the large majority of pigs kept wholly indoors on concrete or slats, it has been widely assumed that intestinal worms would no longer be a health hazard. It has, however, become increasingly clear that while clinical illness from internal parasites is unusual, there is widespread sub-clinical infection with the large intestinal worm *Oesophagostomum dentatum*, as the most important species in breeding stock.

Sows having a moderate infection of about 4000 worms (which in the U.K. can be as many as 20 per cent of the breeding herd) have been shown, in comparison with worm-free sows given the same diet and housing conditions, to lose over 13 kg from service to weaning their litters, in contrast to a gain of over 11 kg during this time by worm-free sows. Less piglets were born and the worm burden of the sows was apparently responsible for lower birth weights of piglets and weaning weights in spite of eating more creep feed. The administration of an appropriate anthelmintic before mating should be routine procedure.

Organization of Preventive Medicine for Pig Units

The usefulness of a preventive medicine service for a pig unit, as indeed it does for farmers engaged in other types of animal production, depends to a large extent upon the care with which each visit is organized the expertise displayed in examining and deciding what action should be taken in each section of the enterprise, and the completeness of the record made subsequently.

Muirhead is a veterinarian with much experience of this type of work, and in a valuable account that gives guidelines in the form of flow diagrams, he has explained how successful such a service can be. Based upon his own records, covering 20 units of varying size, it is shown that great savings can emerge for the farmer for as little as 0.17 per cent per year of the original capital investment per 100 sows.

Vices

With the move, for reasons of economy, to varying degree of crowding in unstrawed concrete or slatted fattening pens, there has been a marked increase in tail and ear-biting. This frequently starts shortly after weaning at 6-7 weeks. In some mixed groups it can become so bad as to cause deaths, both directly and from abscess formation in the spinal column. It has been shown to be worse amongst males. A common method of trying to prevent this vice is to remove the distal part of the tail in the very young pig. But a more certain way is to avoid overcrowding and give the pigs straw bedding.

Sows kept in stalls will sometimes develop the habit of gnawing at the front bars as well as stereotyped head weaving. The only way to prevent this behaviour is to give the sows more room for movement. Again the provision of ample straw bedding will prove helpful.

Poultry

Environment and Nutrition

The modern strains of layers and broilers are now to be found all over the world. They have been bred to convert food energy and protein into eggs and meat at a high level of efficiency. An interference with their rapid growth at the chick or rearer stage is likely to lead to serious financial loss. This means that there should be continuous monitoring of the environment to see that the temperature is kept as close to the optimum for the age of bird as possible, and that the ingredients of the ration are of high quality, as well as in constant and correct proportion. In general, ad lib feeding is practised. The ready availability of ample supplies of pure water has been emphasized.

The commonest cause of trouble is *epizootic disease*, with causative agents and severity varying from one country to another. Not so long ago, it would probably have been true to say that the commonest cause of loss in every country was Marek's Disease, in both its visceral and nervous forms. In the 1960s an effective vaccine was produced, following the identification of the virus that causes the disease. This must be administered immediately following hatching and has now greatly diminished the incidence of this disease.

Virus Diseases

Newcastle Disease (fowl pest) is still the cause of serious loss in some countries, although there are effective vaccines to afford protection. Where the disease is enzootic the vaccine must be given one way or another - intraocularly, parenterally, in the drinking water or as an aerosol spray - on a number of occasions, depending upon the time and severity of the previous outbreak and level of immunity in the parent stock. With young birds, for example, in broilers the first vaccination of an inactivated strain may be at 7 days, 14 days or even 3 weeks, with a second dose given 3 weeks later. Laying birds are vaccinated four times, twice as above, then the third at 6 weeks and the fourth at 16 weeks. For additional protection, they may be given a live vaccine in the drinking water or as a spray every 3-4 months.

Contagious bronchitis, infectious bursitis and epidemic tremor are other virus diseases with wide distribution, while others, like laryngotracheitis, may be a problem in particular areas, as it has been in Australia. Vaccines with varying efficacy have been produced against all these diseases.

Bacterial Disease

The serious bacterial diseases of poultry, such as Pullorum Disease, have been virtually eliminated from the main producing countries through testing and vaccination. There are poultry-adapted strains of *Escherichia coli* that cause sporadic cases of septicaemia, and the avian strain of the tubercle bacillus is to be found where large numbers of layers are kept for a second year.

Hygiene Measures

In addition to dependence upon vaccines to help control infection, there must also be use of the well-tried method of keeping disease at a low level by seeing that the poultry house is kept clear of birds for 10-14 days after each batch of broilers is marketed and when hens have finished laying. First, all litter must be removed and the interior scrubbed down with brushes or high pressure hose, with subsequent application of an effective disinfectant. Laying cages have to be taken apart and thoroughly cleaned. Then the house is left vacant for the specified time.

Not only should all these things be routinely carried out, but above and beyond them all, the best overall safeguard is to have well-trained dedicated stockmen in charge of the birds. The regular and skilled attention this person can give is the best way of keeping losses to a minimum.

11

Animal Welfare

In the U.K., a glance backwards at the way public concern has slowly moved the legislature over matters of animal welfare will show us that Parliament has become more responsive in these latter days to the public's disquiet. No doubt this is in large part due to the power of the press, broadcasting and television to inform with speed and give all too often a highly coloured account to all parts of society.

It is certain that similar forces are at work in other developed countries. Within the European Economic Community there are new mechanisms, such as the expert committees set up by the Commission, whereby the collective will of member nations in this, as in other fields, can be expressed by means of special Directives. These have the force of law and their provisions incorporated into the national legislation of individual states. The Council of Europe similarly has concerned itself with animal welfare. In 1978, it sent to member governments a Convention for the Protection of Animals kept for Farming Purposes, which was ratified by Britain in January 1979.

A brief glance at the history of animal welfare legislation in the U.K. shows us how slowly the public conscience has become sensitive on this topic. Even by the middle of the last century it is recorded that there were over 200 000 dead larks being delivered each year to the Leadenhall Market for the tables of the wealthy, and even as late as 1870 it is recorded that, of a hundred horses observed in Hyde Park, thirty were lame, while of 607 observed working in the streets of Edinburgh, 171 were unfit for the same reason.

The formation in 1824 of the Royal Society for the Prevention of Cruelty to Animals undoubtedly was an important event since it was the efforts of this Society which managed to have the Cruelty to Animals Act of 1835 passed, making it an offence to beat, abuse, overdrive or torture or to fail to provide any confined animal with food and water. This act also extended to the whole country by the provisions of the *Metropolitan Police Act* of 1834 which prohibited the keeping of a place for the baiting or fighting of bulls, dogs, badgers, bears, cocks or any other kind of animal whether domestic or wild. In an effort to prevent worn-out horses being worked or sold by knackers, it required all knackers to slaughter within three days all the horses delivered to them.

Nevertheless, it was not until 1911 that the various pieces of legislation that had slowly been passed during the preceding century were brought together in the important *Protection of Animals Act* of that year. This is the Act under which it is possible to take action for the prevention of unnecessary suffering of domestic animals and any wild animals in captivity. It is under this Act that a great deal of suffering has been ended or prevented, in many instances with the help of the Inspectors employed by the R.S.P.C.A. It must be remembered, however, that it is necessary for cruelty of any sort to be witnessed and thus it gives no protection to farm animals kept out of sight of the public gaze.

Pit Ponies

Perhaps it may now seem a small matter, but at the turn of the century very large numbers of ponies were used to work the shafts of our coal mines. There were some Coal Mine Regulations introduced in 1887 which aimed to provide for the general welfare of ponies and for the submission of reports by Inspectors who were appointed by the Secretary of State, but it was not until the *Coal Mines Act* of 1911 that there were stricter measures brought in. But as with horses used for transport on the roads, it was the coming of mechanization which finally saved large numbers of ponies from having to work underground. Even in the Scottish mines in 1913 there were nearly 6000 ponies being kept underground, but by 1963 this had dropped to only 15.

Experimental Animals

A measure of the greatest importance that aimed at protecting any sort of animal to be used in scientific experimentation was the *Cruelty to Animals Act* 1876. This is still in operation and makes it an offence to perform on a living animal any experiment calculated to give pain, except on satisfying a number of very rigid restrictions, including the fact that the person performing an operation must be licensed and must be able to show that the experiment is likely to advance physiological knowledge or knowledge which will be useful in the saving or prolonging of life or alleviating suffering.

Laws to Prevent Cruelty

Since the passing of the Protection of Animals Act 1911, there have been a number of amendments to the main Act and various other Acts concerning the prevention of cruelty to dogs and the use of anaesthetics to perform various operations such as castration or the docking of tails. It means that neither male cattle nor sheep can be castrated without an anaesthetic after they reach the age of three months, two months for goats and pigs. If a rubber ring is to be used for castration then this must be used within the first week of life, or it has to be done with an anaesthetic.

Amongst the eleven other Acts which seek to protect animals that are in contact with the public are the *Performing Animals (Regulations) Act* 1925 which controls the training and exhibition of a performing animal, whether it is of a wild or domestic nature, other than invertebrates and animals trained or exhibited by bona fide military police or used for agricultural or sporting purposes. The *Animal (Cruel Poisons) Act* 1962 prohibits the killing of any mammal by means of cruel poisons. The Regulations under this Act cover the use of strychnine, yellow phosphorus or red squill.

So far as the well being of animals kept for commercial purposes is concerned, there are four important Acts here which are concerned with the sale of pet animals, the keeping of boarding establishments for animals, the keeping of horse-riding establishments and the breeding of dogs. In effect, it means that all these activities have to be licensed annually by the appropriate Local Authority. With animal boarding establishments and horse-riding establishments, the premises have to be inspected by veterinary surgeons. In 1971, the *Animal Act* was passed which covers four important matters concerning animals which may be kept on farms. It sets out the strict liability for damage done by animals; it provides for the detention and sale of trespassing livestock, and the liability of the owner of the animal for livestock straying on to the highway, and there is provision for the protection of livestock against dogs. This last section makes the keeper of a dog strictly liable if the dog kills or causes injury to any livestock.

Welfare of Farm Animals

It was the publication in 1964 of a book entitled *Animal Machines* by Ruth Harrison that provoked sufficient public feeling to induce the then Minister of Agriculture to form a Committee of Enquiry 'to examine the conditions in which livestock are kept under systems of intensive husbandry and to advise whether standards ought to be set in the interest of their welfare, and if so what they should be'.

It reported in the following year, in the meantime becoming known as the Brambell Committee, after its Chairman Professor Rogers Brambell, a distinguished zoologist. From that report came further legislation of the greatest importance to the welfare of farm animals. But this did not happen until 1968 for the Government was very cautious over the matter. The Brambell Committee's report however has continued to have good deal of relevance for several reasons. It was the first public enquiry into intensive husbandry methods. It surveyed those methods with different classes of farm animals and was able to show how serious was the absence of scientific data, in terms of being able to measure the degree of stress to which different species were submitted in these new intensive systems of husbandry. It also devoted a chapter of its report to the problem of stockmanship, recognizing that at the very heart of this matter is the skill and devotion of the animal attendant. It made an important statement on what should be the minimum objectives in any intensive farm animal husbandry system. And finally it proposed two things: first that there should be a Farm Animal Welfare Advisory Committee to continue surveillance of the field and be able to advise the Minister of Agriculture, and, second, that a new Act was needed to give a fuller definition of suffering and to allow regulations to be made.

Codes of Practice

In the decade that has passed since this committee was finally set up by the Government, it has been responsible for the publication of what have been called 'Codes of Practice', dealing, respectively, with Cattle, Pigs, Fowls, Turkeys and Sheep. Each of these small booklets contain a series of recommendations to farmers, most of them in very general terms. Indeed, the full title of each booklet emphasizes what they are—*The Codes of Recommendations for the Welfare of Livestock*. If a livestock farmer does not carry out any of these recommendations, they do not have any force of law. The only thing is that if it is known by the Government veterinary inspectorate that he has been persisting in a practice against the code after being advised to alter his system of management, it will be counted against him if a prosecution is brought. Each of the codes so far published follow the same pattern. New revised editions are at present being prepared.

They begin with a preface which it points out is not part of the code but is intended to show on what principles it has been based. It begins by quoting from the Act to explain the status of the codes, then states what are the basic requirements. These are in fact largely those that were stated in the Brambell report.

They include the need to have fresh water and nutritionally adequate food available as required by the livestock involved; the provision of adequate ventilation and suitable environmental temperature; adequate freedom of movement and ability to stretch limbs; sufficient light for satisfactory inspection; the rapid diagnosis and treatment of injury and disease; emergency provision in the event of breakdown of essential mechanical equipment; flooring which neither harms nor causes undue strain; the avoidance of unnecessary multilation.

The Codes themselves have short sections under such headings as 'Housing', 'Space Allowance' -where, however, there may be no actual dimensions suggested—'Food and Water' and 'Management'. It has been a common criticism of these official recommendations that they are too vague, for who is to decide what is 'sufficient trough space' or 'sufficient sideways movement'; or 'adequate ventilation'? What emerges is that the welfare of livestock in the last analysis mainly depends upon the competence and conscience of the stockman who is in day-to-day charge.

Need for research

As our understanding of disease prevention extends, so should we be seeking to extend our understanding of the ethology of farm and other domestic animals. So far, it has proved difficult to define satisfactory criteria that measure the degree of stress to which stock are submitted under modern systems of husbandry. The wide differences that are obvious when different breeds and strains within breeds are submitted to the same environmental conditions, indicates how important is the genetic make-up of each animal. It is significant when we contemplate such a dearth of knowledge in this area that the Brambell committee felt it necessary to devote a whole chapter to the subject of the role of the stockman and the need for their careful selection and training.

Farm Animal Welfare Council

The continuing and mounting concern of the public in animal welfare may be seen in the move made in 1979, shortly after a change of government in the U.K. The Farm Animal Welfare Advisory Committee was changed and its membership enlarged to become the

Farm Animal Welfare Council. Its terms of reference were extended to include the welfare of farm animals at the place of slaughter, including the handling of animals before slaughter and methods of slaughter. So the revised terms of reference were to keep under review the welfare of farm animals (1) on agricultural land, (2) at markets, (3) in transit, (4) at the place of slaughter, and to advise the Minister of Agriculture of any legislative or other changes that may be necessary.

THE LAW

It was another of the Brambell Committee's recommendations that there be an Act of Parliament under which authorized persons would have the right of entry to inspect intensive livestock enterprises. In 1968, The *Agriculture (Miscellaneous Provisions) Act* became law. It states that 'any person who causes unnecessary distress to any livestock for the time being situated on agricultural land and under his control or permits any such livestock to suffer any such pain or distress of which he knows or may reasonably be expected to know shall be guilty of an offence under this section'. There are two ways in which farmers are to be encouraged not to become liable under the Act. The first of these is by the making of regulations which are binding and may state in precise detail matters concerning the nature of accommodation for animals, anything to do with 'the provision of balanced diets and for prohibiting or regulating the use of any substance as food for livestock and the import and supply of any substance intended for use as food for livestock'. The third category gives power to regulate the bleeding, multilation, marking 'or interfering with the capacity of livestock to smell, see, hear, emit sound or exercise any other faculty'.

Until 1979, the only regulations promulgated under the Act were made in 1974 and come under this third category. It is an offence in the U.K. to dock the tails of pigs over 7 days old except when the operation is performed by a veterinarian on health grounds or to prevent injury from tail-biting. In such instances, an anaesthetic must be used.

With cattle, it is prohibited to dock their tails. With poultry, it is forbidden to castrate male birds surgically; birds may not be pinioned (i.e. the brachial ligaments may not be severed) and blinkers may not be fitted by a method involving multilation of the nasal septum. It will be seen at once that, apart from the docking of pig's tails, the practices prohibited have very rarely been undertaken by farmers in the U.K.

In order to be able to ratify the Council of Europe's Convention on the protection of animals kept for farming purposes, a Regulation came into force in January 1979 that makes it obligatory for all livestock and automatic equipment in intensive units to be inspected daily. Where the animals are found not to be in a state of well being, measures must immediately be taken to safeguard the livestock from suffering unnecessary pain or unnecessary distress. Any defect in automatic equipment shall be rectified forthwith, or measures taken to safeguard the stock until it is rectified.

Along with the right of entry, a duly authorized veterinarian may examine any livestock he finds on the land and apply to and take from the livestock such tests and samples that he considers appropriate. Thus, there is available in this country a powerful piece of legislation in the interests of intensively kept, or any other form of livestock husbandry. The overwhelming difficulty at the present time is to be able to apply objective criteria that can define the degree of stress or discomfort to which animals or birds are being subjected.

Increasing Public Concern

The fat that the Council of Europe's Convention aimed at protecting farm livestock could be ratified so speedily by six of the nine E.E.C. countries and three others (Sweden, Norway and Cyprus) clearly indicates the widespread concern amongst the public about animal welfare.

Sweden with its Code of Statutes 1974 established close control over animal accommodation and forbids permanent tethering of pigs and beak-trimming in poultry. Norway, under its Welfare of Animals Act 1974, similarly forbids the latter practice. It also forbids the insertion of rings into pigs' snouts, as does Denmark. Switzerland, in an Ordinance in 1979, has proceeded to lay down limitations on the numbers of various types of animals that may be kept in a building.

The difficulties of funding and conducting research in this field are indicated by the fact that in the period 1965–1972 there had been a reduction by one-half in the number of full papers published that dealt with veterinary ethology appearing in the British scientific literature each year.

The general public is, by the aid of the press, radio and television, now much better informed about some of the more restrictive practices commonly used in intensive pig and poultry production units. This probably accounts for a persistent and mounting concern over such things as the keeping of layers in battery cages, of sows in stalls and of the rearing of 'white' veal calves in individual small stalls.

It is not difficult to understand why these methods have become so widely adopted. Any farmer in business to make money finds it hard to resist economic pressures to reduce space allowances as building costs escalate, often at the same time reducing the number of animal attendants because of high wages. Consequently there is likely to be a reduction in the frequency of inspection to permit the discovery of the animal in difficulty or showing the first signs of illness.

Possible Improvements

At the present time a Select Committee of the House of Commons is investigating the possible need for more stringent regulations to promote a greater degree of farm animal welfare. A group of churchmen and distinguished behavioural scientists have published a book. To a nation notoriously obsessed by its concern over pet animals, it is sobering to read, 'the basic welfare needs of the calf or piglet are neither quantitatively nor qualitatively different from those of a puppy or kitten; the welfare of farm animals is assured only so long as the environment provided by man is not outside its ability to adapt without suffering'. This group affirms the standpoint that an acceptable system of husbandry must, as far as possible, be based on permitting all forms of behavioural and physiological potential that are available to the wild or ancestral species This means safeguarding the basic freedoms that were first adumbrated in the Brambell Report and have been adopted in the British Codes of Recommendations for Welfare.

In the absence of agreed objective criteria to measure degrees of distress, and acknowledging that high levels of productivity can be attained under crowded conditions, the only way in which intensive production processes judged on subjective grounds as being wrong can be abolished or altered is by being able to demonstrate that profitability will not be reduced by using an alternative 'acceptable' method.

House of commons committee on agriculture

A newly appointed all-party Select Committee on Agriculture decided in 1980 to investigate animal welfare in poultry, pig and veal calf production. In addition to receiving written statements and taking oral evidence from a large number of farmers, veterinary, Government and welfare organizations, visits were made to a number of intensive units, to research institutes both in the U.K. and abroad. It issued its report in 1981 and made some important recommendations. Amongst those of a general nature were that there should be increased research into behavioural and ethological problems, that the U.K. Government should give 'a strong and sustained lead in the European Community towards increased animal welfare, and that inspection of indoor intensive production units should be inspected at least once annually by the veterinary inspectorate specifically directed towards welfare'.

On veal calves, they urge the Minister to seek European Community agreement to measures that will bring an early end to veal calf rearing in crates. Labelling should be encouraged that makes clear the method of production.

On the keeping of sows, they want efforts to be made to develop alternatives to the close confinement of pregnant sows. Those who use small groups should be given financial grants for such housing, but no grants for sow stalls.

They hope for E.E.C. prohibition of castration for pigs destined for meat and the subsequent banning of imports from sources which infringe this restriction.

For laying hens kept in battery cages, E.E.C. agreement is sought on a minimum standard of 750 cm^2 per bird, but the Minister should refuse anything less than 500 m^2 per bird. Further, it is urged that agreement in the Community should be sought for a statement that 'after, say, five years egg production will be limited to approved methods which will not include battery cages in their present form'. Again, it is recognized that imports from countries that do not have equivalent restrictions will be stopped. Beak trimming should be forbidden under Regulation except under veterinary supervision 'as a last resort when it is clear that more suffering would be caused if it were not done'.

Hens in battery cages

So far as the battery hen is concerned, there seems at the moment to be no certain way of giving it a better, or more interesting and satisfying existence. Modifications of the standard cage have been tried, sometimes called the 'get-away' cage. The main characteristics of this, as in other experimental cages with a similar purpose, is to provide nesting and perching areas, as well as sufficient space for feeding, all in the same floor area as the standard cage. This latter because it is realized that unless approximately the same total number of layers can be kept in existing houses, the chances of having the new type of cage accepted by the industry would be diminished. On the other hand, there would not be the same objection to making the height of a cage somewhat greater. The purpose of having a nesting area with suitable litter is because research has shown that the greatest recurring frustration for the laying hen in a battery cage is the lack of a place in which to go to lay her egg. Comparative trials with similar hens in the same house kept either in the get-away cages or in conventional cages have shown that, while the total number of eggs was not greater, the egg weight over a period of many weeks was slightly increased, the extent of feather pecking was much less and the condition of the birds' feet was better in the get-away cages. Serious commercial drawbacks are the soiling of eggs with faeces and the greater labour requirement involved in egg-collecting and grading.

The aviary system of housing layers is a development of the 'get-away' cage. Still only in an experimental stage, it may in time permit general abandonment of the present type of battery cage. At Gleadthorpe Experimental Husbandry Farm, scientists of the advisory service of the U.K. Ministry of Agriculture have begun observations on a new type of aviary house.

Cambridge straw-yard trial

Over five years, workers at the Cambridge Veterinary School have been able to show that hens kept in cheaply covered straw yards have laid as many eggs with a slightly lower feed requirement as similar hens kept in battery cages in an environment-controlled house. This is the second method that has sought to avoid the objectionable features of both cages and the deep litter system for layers. The mortality rate was 10 per cent in the straw yard, compared with 12 per cent in the battery birds. It is claimed by Sainsbury (1981) that there are the following advantages in this method: the building is cheap to construct and adaptable; running costs are low, since there are no fans and a minimum of lighting; there is nothing mechanical to go wrong; it converts straw into valuable manure; the quality of the eggs is good and the straw yard eggs are richer in colour and in vitamin B_{12}. On the other hand more space is required, there needs to be a supply of straw and, since feeding and egg collection is by hand, there is presumably a higher labour charge. It is, in other words, not a system easily adapted to the very large-scale units that nowadays produce most of our eggs.

Sows

With the widely adopted sow stalls, using either a tether or back rail to keep each animal in place, there is certainly avoidance of the fighting that can occur when sows are kept together. If feeding is by hand, individual feed intake can also be monitored. Nevertheless, such continuous close confinement frustrates many of the animal's natural movements and desires. These frustrations lead to stereotyped behaviour such as head-shaking and gnawing at the bars of the stall.

While the housing of pregnant sows in small groups of similar age and breed in a straw yard with individual feeders, is likely to achieve satisfactory levels of welfare as well as production, demands on space and labour are increased. There is a need at the present time for critical testing of alternate systems of accommodating pregnant sows with their welfare in mind.

Perhaps some answers will emerge from two interesting research programmes at present being pursued within the U.K., both designed to extend our knowledge about the range and variation of the modern pig's daily needs and, by constructing accommodation that can satisfy these to a large degree, prevent the distress that most of the present housing systems inevitably cause. One, centred in Edinburgh, has made detailed observations upon a group of sows and boars kept in a large wooded enclosure without housing. With knowledge of the main behavioural patterns of family groups, they have made this the basis of a new method of housing. This consists of an area covered in peat to permit rooting with another activity or resting area next to it that is strawed and a feeding area with individual stalls. In such a system the labour input is a little higher, just as the space allowance for each sow, 25.3 m^2 compared to 16.3 m^2 in conventional farrowing-to-finishing buildings, is distinctly greater. It has been shown, however, that it is possible to rear and fatten 20 pigs from two litters each year and that the pigs reach bacon weight at less than 150 days.

The other study in depth upon pig behaviour that may lead to modification of current husbandry and housing practice is being undertaken at the Agricultural Research Council's Institute of Animal Physiology at Babraham. Here is a series of experiments in operant conditioning are helping to pinpoint many physical preferences over such factors as the house temperature, the type of flooring and light intensity. A variety of feeding patterns were imposed and, as with all other variables, there was a wide range of individuality exhibited. Above all, the pig's supreme gift of adaptability was repeatedly emphasized.

Multilations

Tail-biting in young pigs can be a serious problem in some piggeries, where there is overcrowded accommodation, without bedding, while a variable or deficient diet are believed to be contributing factors.

For a unit to feel compelled to dock all the tails of young pigs to avoid tail-biting and its often associated ear-biting, is an admission of there being an unsatisfactory system in operation. When weaners are kept in open straw yards there is no problem from this vice.

It has been known for some time that with modern early maturing breeds given optimum feeding, there is no justification for the castration of young male pigs, ram lambs or bull calves.

While the pheromone produced by the boar can be detected by a few people in a small proportion of baconers that have not been castrated, this is destroyed on cooking and increasingly it is expected that boars will not be castrated if they are to be slaughtered as porkers or young baconers.

In some European countries, the castration of bull calves has been stopped for many years, and so long as the accommodation to rear bull calves in groups is adequate, there is no reason to castrate male calves. It is similar situation with ram lambs of modern hybrid breeds of sheep that have been brought quickly to slaughter weight. There is still some prejudice amongst butchers in some countries, but this is likely to diminish with time.

There is no doubt at all that the better food conversion of entire male animals with a larger proportion of lean to fat are attributes that can be usefully exploited.

White Veal Calf

It has been the practice in several European countries for many years to take calves from dairy cows that have been removed from their dams after 24–36 hours and rear them separately on a milk replacer diet in crates, generally made of wood. Instead of weaning them at the age of 5–8 weeks, which is the normal practice with calves to be used for dairy replacement or for later fattening for beef, these calves remain in crates and are given a high fat milk replacer without any forage or concentrate feeds. They are often given antibiotics in their feed in an attempt to reduce intestinal or pulmonary infections, and thus are able to make high weight gains until the age of 14–16 weeks. At this stage they have fully occupied their small crates, and of course have been for most of this time unable to turn round, only being permitted to get up and lie down. At the age of 14–16 weeks they are taken out and slaughtered. In order to reduce worry from flies and also to keep the animals quiet, they have commonly been kept in complete or semi-darkness for most of this time. It is claimed by producers undertaking this system that the animals are carefully tended, they are protected from extremes of weather for their environment is kept at a constant level of temperature and humidity; they are kept free from disease and they are in fact better cared for than calves under some other systems.

The facts of the matter are, however, that the offspring of ruminants which are herd animals, have a natural proclivity to play together and certainly proper feeding would encourage them to take some forage after about a month kept on a milk diet so that their rumens may begin normal development.

Mounting public concern about this system has prompted a few producers in Continental countries and in the U.K. to look critically at some of these assumptions to see whether it is possible to rear veal calves to a satisfactory weight at 14–16 weeks, allowing them to have more liberty, keeping them in small groups with some straw bedding.

Some interesting recent results obtained by Webster shows that it is entirely possible to produce acceptable calves under a system which does not depend upon keeping the calves in individual crates nor even in an environment-controlled building. This system does however depend upon very careful feeding, disease control and a high level of stockmanship.

The evidence is sufficiently convincing to have induced many commercial enterprises in the U.K. to change to the straw-yard system.

Export of Animals

Another matter that has caused a good deal of public concern over welfare aspects concerns the conditions under which live animals intended for slaughter are exported from this country. There are two classes of stock involved here. One, and the smaller, concerns the export of valuable stud stock to many countries. These animals are nowadays nearly always sent by air and there has been an agreement amongst the International Air Transport Authorities on the requirements for stock to be carried by plane. The U.K. State Veterinary Service has been much involved in the setting out of conditions for the export of animals by air and it can be stated now that the conditions are as satisfactory as they can be made with present transport planes, loading and unloading facilities.

The legal requirements are embodied in the Transit of Animals (General) Order 1973. There is at present a comprehensive Code of Practice for the Transport by Air of Farm Livestock and Horses being finalized. It has some 59 sections and will further safeguard the welfare of all types of animals being transported by plane.

Much larger number of stock are sent by sea, either store cattle exported from Eire for fattening in this country or sheep and calves sent to the Continent for fattening in France, Italy or Germany. Some years ago publicity was given to accounts of long journeys across Europe made by lorry loads of sheep and calves without adequate rest periods or proper attention to their need for water or food. So great was the public reaction to these accounts that for a period in 1973–1974, the export of live animals from the U.K. was stopped. Following the report of the O'Brien Committee in March 1974 recommending that, with appropriate safeguards, the transport of these animals could be made satisfactorily, it has been resumed. The conditions under which animals could be exported from ports within this country were significantly tightened as a result of the Exported Animals Protection Order 1964. This made it necessary for more information to be provided on licences being issued by the Ministry of Agriculture for stock to be exported, it tightened up regulations so far as the unloading of stock at lairages at the designated ports before loading on to ships and, by increasing the inspectorate for this trade, has helped to see that suitable transport is used for animals being sent by ship to Europe.

A further valuable increase in the safeguarding of the Welfare of animals being transported within all E.E.C. countries and to other countries has been the bringing into force of the *E.E.C. Directive No. 77/489/EEC*. This provides that there must be an international animal transport certificate with every consignment of animals and it sets out in detail the nature of the consignment, its destination and the route that it will follow; this has to be signed by an official veterinarian and when the animals are loaded there is a separate section where this is stated and signed and at each control post that the consignment passes through there is provision for further noting of date and time. If there is any change of destination, this has to be specified, as does the feeding and watering arrangements, with a final attestation at their destination. The Directive itself sets out the need for care in the handling of the animals and that the journey must be completed as quickly as possible. It would seem, therefore, that if the legal framework is properly operated by conscientious people, there is now some safeguarding of the welfare of transported animals within the E.E.C. countries.

All this of course does not answer the objections of those who say that animals that are to be slaughtered should only be transported the shortest distance within their own country and that only carcasses should be exported. It may also be fairly stated that while livestock would presumably not be loaded into cross-channel boats when the seas are rough, no one can be certain that there will not be considerable discomfort to animals when conditions are not particularly smooth. And it cannot also be gainsaid that a lorry transporting animals over long distances by road will sometimes encounter unexpected delays when the stock may well be submitted to some distress.

Since the Transit of Animals (Road and Rail) Order 1975 and its Amendment Order were made, the conditions under which animals may be transported within the U.K. have been more carefully regulated. The provisions within six older orders are replaced so that there are now much more detailed requirements for the construction and maintenance of vehicles and receptacles used to transport animals, and for safeguarding their welfare during loading, carriage and unloading.

It was nevertheless felt at a conference of scientists, members of the meat trade and processors, representative of welfare organizations and ethologists, that much more needs to be done to minimize distress in the handling of animals before loading, during transport and on arrival at market or abbatoir.

Welfare of Animals being Slaughtered

Within the U.K., as no doubt in other western countries, there has been a steady tightening of the requirements concerning the transport and handling of animals being taken to abattoirs. The main requirements in the U.K. are covered by the Slaughter of Animals (Prevention of cruelty) Regulations 1938, which were in part further strengthened by the *Slaughterhouse Act* 1974. The structure and organization of lairage accommodation with requirements for the protection of stock from sun and adverse weather and separate accommodation for horned and fractious animals are all covered. There are stated requirements for feeding and watering of animals and if any are ill or in pain they must be slaughtered as soon as possible. All animals of course must be rendered unconscious before slaughter except for those being dealt with by ritual slaughter under the Jewish and Moslem requirements.

The Ammerdown Group (1980) were in no doubt that there was room for much improvement in the present routines followed in many slaughterhouses. They made a series of recommendations whose implementation would without doubt help to reduce the suffering of all types of animals used for meat. Especially important was their belief that 'a small independent co-ordinating centre should be established as soon as possible to act as a focal point for all those concerned with the transport and slaughter of farm animals'.

The Slaughter of Poultry (Human Conditions) Regulations 1971 have helped to ensure humane conditions for turkeys and domestic fowls being slaughtered for human consumption. They deal with the way that the birds must be kept before slaughter and any bird in pain as a result of injury or other cause must be slaughtered immediately. The conditions under which the birds are actually slaughtered are also laid down.

It is a requirement of the regulations that all birds must be rendered unconscious before exsanguination, accomplished by machine or by hand. The approved method at present is by electric stunning, achieved in most poultry abattoirs by the inverted shackled birds having their heads immersed in a water trough through which an electric current from the mains is passed at 90 volts. This is effective so long as every bird has its head immersed. Very occasionally this, for a variety of reasons, does not occur. There has also been concern expressed by veterinarians at even well-run plants because of the erroneous belief amongst managers of such places that dead birds will not bleed, there is the tendency to run the stunning bath at a lower voltage. Careful observation has also revealed that if the severance of the carotid arteries is not achieved the minimum bleeding time of 90 seconds may not be sufficient before the birds enter the scalding tank.

Conclusion

Adequate legislation is the necessary framework within which actions believed by most citizens to be desirable can be achieved and undesirable ones prevented. There have been notable and praiseworthy advances in recent years towards safeguarding the welfare of farm and domestic animals. It is seen in context of the general public's concern and the strengthened laws covering so many aspects of the subject. But laws are like old-fashioned forts, to be effective they must be well manned. We should remember that most farmers are activated by the same abhorrence of causing unnecessary suffering as are other members of the public. Knowledgeable and humane farmers and animals attendants advised by an informed veterinary service and buttressed by enlightened legislation strictly enforced are necessary to keep avoidable distress and illness at a minimum, not only in the intensive husbandry units that have become a feature of the present farming scene, but in all animal production systems.

INDEX

A

Agmark, 229
Agriculture (Miscellaneous Provisions) Act, 287
Animal (Cruel Poisons) Act, 284
Animal Act, 284
Animal Machines, 284
Aspergillus, 213

B

Back-crossing, 150
Biotin, 115
Blemish, 251
Blindness, 253
Bog spavin, 252
Bone spavin, 252
Bordetella, 277
Bos indicus, 144
Bovidae, 231
Bovine virus diarrhoea, 213
Breed societies, 138
Brucella abortus, 211
Brucella canis, 215
Brucelıa ovis, 189
Brucella suis, 214
Bubalus bubalis, 232

C

C. foetus, 211
C. foetus intestinalis, 213
C. foetus venerealis, 211
C. pyogenes, 241
Calf-rearing, 49
Cambridge, 147
Campylobacter vibrio foetus, 211
Capped hock, 253
Carbohydrates, 132
Chlamydia, 213
Coal Mines Act, 283
Cocked ankle, 253
Coliforms, 241
Collecting yard, 54
Colostrum, 164, 206
Comtemporary comparison test, 149
Coopworth, 148
Copper, 114
Cotton mules, 254
Cotyledons, 198
Cruelty to Animals Act, 283
Curb, 252

D

Damline, 147
Darjeeling bristles, 229

Degradability, 78
Deshi, 233
Desi bristles, 229
Dictyocaulus viviparus, 259
Difficulty of detecting oestrus, 270
Digestible energy, 82
Draft mules, 254
Drafting race, 55
Dry matter, 70

E

Electro-ejaculator, 191
Energy, 69
Epizootic disease, 280
Escherichia coli, 258, 281

F

Farm mules, 254
Fat, 131
Fat soluble, 72
Fattening, 49
Finishing, 49
First stage of labour, 205
Fistula, 252
Follicle-stimulating hormone, 200
Follow-on, 49
Foot bath, 55
Forcing pen, 54
Fusiformis necrophorus, 263, 275

G

Gene, 138
Gromark, 148
Grower pens, 49

H

Haemagglutinin inhibition test, 202
Haemophilus equigenitalis, 210
Heaves, 253
Herringbone, 42
Heterosis, 152

I

Improved contemporary comparison, 149
Inadequate nutrition, 270
Inbreeding, 145
Intestinalis, 211

J

Jaffarabadi, 235

K

Karyotype, 150
Klebsiella aerogenes, 210

L

Lactating female, 132
Line-breeding, 146
Loose boxes, 32
Luteinizing-hormone, 200

M

Maintenance, 81
Melilotus alubs, 73
Metabolizable energy, 86
Metropolitan Police Act, 283
Minerals, 69
Mining mules, 254
Mitosis, 139
Mouse test, 202
Murrah, 233
Mycoplasma bovigenitalium, 212
Mycoplasma hypopneumoniae, 277

N

Nature, 81
Net energy, 82
Nili, 234

O

Oesophagostomum dentatum, 279
Oestrogen, 200
Optimum weight, 102
Ostertagia, 259, 260
Ovarian dysfunction, 270

P

Per rectum, 203, 273
Perendale, 147

Performing Animals (Regulations) Act, 284
Piggery, 222
Placenta, 199
Poll evil, 252
Pregnancy, 132
Progesterone, 200
Protection of Animals Act, 283
Protein, 69, 131
Pseudomonas aeruginosa, 241
Pyrethrins, 39

R

Racehorses, 85
Ravi, 234
Ribosomes, 138
Ringbone, 252
Rumination, 75

S

S. abortus-ovis, 214
S. penortha, 275
Salmonella, 214
Salmonella abortus equi, 211
Shearing shed, 55
Shute, 42
Sidebone, 252
Siphona exigua, 245
Slaughterhouse Act, 296
Splint, 252
Stalls, 32
Standing test, 196
Stifled, 252
Stomoxys calcitrans, 245
Streptococci, 241
Streptococcus lactis, 80
Stringhalt, 252
Sugar mules, 254
Surti, 234

T

Tandem, 42
Tarai, 236
Teat dipping, 262
Thiamin, 73
Thoropin, 252
Time, 81
Toda, 239
Toxoplasma gondii, 214, 215
Trichomonas foetus, 213
Trichostrongylus, 259
Trypanosoma equiperdum, 211

U

Unsoundness, 251

V

Villi, 198
Vitamins, 70

W

Water, 70
Water soluble, 72

Z

Zebus, 144